Palgrave Studies in the
and its Su⌐⌐⌐⌐⌐⌐

MW00835556

Series Editors
Calvin Mercer
East Carolina University
Greenville, NC, USA

Steve Fuller
Department of Sociology
University of Warwick
Coventry, UK

Humanity is at a crossroads in its history, precariously poised between mastery and extinction. The fast-developing array of human enhancement therapies and technologies (e.g., genetic engineering, information technology, regenerative medicine, robotics, and nanotechnology) are increasingly impacting our lives and our future. The most ardent advocates believe that some of these developments could permit humans to take control of their own evolution and alter human nature and the human condition in fundamental ways, perhaps to an extent that we arrive at the "posthuman", the "successor" of humanity. This series brings together research from a variety of fields to consider the economic, ethical, legal, political, psychological, religious, social, and other implications of cutting-edge science and technology. The series as a whole does not advocate any particular position on these matters. Rather, it provides a forum for experts to wrestle with the far-reaching implications of the enhancement technologies of our day. The time is ripe for forwarding this conversation among academics, public policy experts, and the general public. For more information on Palgrave Studies in the Future of Humanity and its Successors, please contact Phil Getz, Editor, Religion & Philosophy: phil.getz@palgrave-usa.com.

More information about this series at
http://www.palgrave.com/gp/series/14587

Adam Briggle

Thinking Through Climate Change

A Philosophy of Energy in the Anthropocene

Adam Briggle
Department of Philosophy and Religion
University of North Texas
Denton, TX, USA

Palgrave Studies in the Future of Humanity and its Successors
ISBN 978-3-030-53586-5   ISBN 978-3-030-53587-2   (eBook)
https://doi.org/10.1007/978-3-030-53587-2

This Palgrave Macmillan imprint is published by the registered company Springer Nature Switzerland AG.
The registered company address is: Gewerbestrasse 11, 6330 Cham, Switzerland

[I]f the choice is to be good, the reasoning must be true and the desire correct.
—Aristotle on "Intellectual Virtue" in *Nicomachean Ethics*

Measure is alien to us; let us own it; our thrill is the thrill of the infinite, the unmeasured. Like a rider on a steed that flies forward, we drop the reins before the infinite, we modern men, like semi-barbarians – and reach our *bliss only where we are most*—in danger.
—Friedrich Nietzsche on "Our Virtues" in *Beyond Good and Evil*

Preface

The Earth's climate has always been changing. On my office wall, I have a picture of me with my children Max and Lulu on the North Rim of the Grand Canyon. The Colorado River has cut through layers of rock: Coconino, Toroweap, Redwall…all the way down to the Vishnu Basement. To run your eyes a mile down to the thin ribbon of water is to travel back in time 2 billion years. The Grand Canyon is pure movement—crashing tectonic plates, exploding volcanoes, the Kaibab Plateau lifting up like a blister, storm clouds gathering, and generations of rivers erasing generations of mountains. The climate roamed from hot to cold, from sea to desert.

The Grand Canyon is a good place to see the climate system at work. The climate system is the interaction of the lithosphere (rocks), cryosphere (ice), atmosphere (air), hydrosphere (water), and biosphere (life). It is how these spheres move around each other and in and out of each other. This restless movement is energy. The climate system is a grand energy flux.

Humans are new to Earth, having only been around for about 200,000 years. To give you some perspective, the *youngest* rocks at the Grand Canyon are about a *thousand times older* than *Homo sapiens*. Yet, we have become agents of planetary change; the energy flux that is the climate is now marked with human fingerprints. How did this happen and what does it mean? Can we learn to become responsible stewards of the climate? Is this too much for us? Have we overreached, or are we only just beginning to realize our potential? *Thinking Through Climate Change* is my attempt to get at these questions.

There are many ways to think through climate change. Energy is the golden thread that I follow. For the past few hundred years, industrialized societies have been pulling carbon out of the lithosphere and putting it into the atmosphere where it traps heat via the greenhouse effect leading to global warming. Much of this extra energy is absorbed by the oceans. Indeed, the extra heat accumulating in the oceans is equivalent to the energy of *five* atomic bomb explosions *every second*.

In 1960, atmospheric CO_2 concentrations stood at 309 ppm. In 2020, CO_2 concentrations topped 415 ppm, higher than at any time in the past 800,000 years. Anthropogenic (human-induced) climate change has current and projected impacts on Earth systems that are wide-ranging. These will in turn impact humans in many ways: food security, access to water, heat stress, extreme weather, disease, migration, war, and more. This is why climate change is a defining issue for the future of humanity and planet Earth.

Climate change is the unintended consequence of the making of a high-energy civilization or petro-culture. Life in the twenty-first century is powered by machines that consume enormous amounts of calories and that move carbon from rock to air. There are lots of good books about the history and politics of these technologies. Yet the machines are not the most important factors. What really matters are the ideas and values behind the machines. That's what this book is mostly about.

Animating our high-energy civilization is the modern concept of energy as a universal currency and as the capacity to do work—a capacity that humans lack and must acquire in order to live a good life. These concepts of energy are claimed by scientists, economists, and engineers nowadays, but they have deep philosophical roots. Indeed, for me, climate change is not just a subject for philosophical reflection; it is the culmination of Western philosophy. Philosophy is a conversation about energies (human, natural, and divine). This conversation birthed the machines that have altered the climate of our planet. Who said philosophy doesn't have an impact?!

I am grateful to everyone who helped me with this project. My wife, Amber, contributed many valuable ideas during our "date night" discussions. I want to thank her for her creativity, support, and encouragement. Max and Lulu are also endless sources of wonder and inspiration. I want to thank them for our adventures in the "Emma Dean," the camper we took to so many National Parks as I thought about this book.

I workshopped many of the ideas in these pages in graduate and undergraduate courses on energy, environment, and climate change. I am indebted to my students for their thoughtful contributions, gentle criticisms, and convivial conversations. Many thanks as well to the following people who all helped me in various ways: Robert-Jan Geerts, Jen Rowland, Terra Rowe, Maggie (Keith) Brown, Robert Frodeman, Glen Miller, Daniel Sarewitz, Mike Cochran, Maya Rao, Dominic Boyer, and Imre Szeman. I am especially grateful for the help and support provided by Carl Mitcham, and I owe a special debt of gratitude to my friend and former PhD student Giovanni Frigo who did so much to advance my thinking about energy.

As a public servant, I am grateful to the residents and taxpayers of the state of Texas and to the University of North Texas. I hope that this book represents a valuable use of the time and freedom afforded to me in my privileged place as a professor at a public university.

When I was despairing about the prospect of ever finding a publisher, Steve Fuller stepped in to save the day. I am especially grateful for his encouragement, energy, ideas, and advice. I am fortunate to consider Steve a friend and mentor. I'd also like to thank the anonymous reviewers for Palgrave Macmillan who offered valuable ideas for improving the text. Finally, I want thank the editors at Palgrave Macmillan, especially Philip Getz and Amy Invernizzi.

Denton, TX, USA   Adam Briggle

Contents

List of Figures

List of Tables

1

Introduction

*The most thought-provoking thing about our thought-provoking time is that
we are still not thinking*
Martin Heidegger *1954*

There is a crack in reality. Our name for it is *energy*. From Heraclitus to
Lao Tzu to Albert Einstein, deep thinking about *energeia*, *qi*, or E has led
to mystery. In *Frankenstein*, Mary Shelley imagined electricity giving rise
to the living-dead. To comprehend the quantum energies at the base of
reality, the physicist Erwin Schrödinger conjured a thought experiment
about a cat that is also simultaneously alive and dead. It is a paradox, a
superposition, a contradiction.

My thesis comes in two parts. Here's the first half: As we build a civilization that uses more and more energy, the crack in reality gets wider and
weirder. Climate change is this growing uncanniness. The ice at the
Earth's poles has long pulsed in and out with the seasons like a pair of
frosty lungs. Scientists have a word for systems that change like the seasons: *stationarity*. It means that the properties that give rise to change are

© The Author(s) 2021
A. Briggle, *Thinking Through Climate Change*, Palgrave Studies in the Future of
Humanity and its Successors, https://doi.org/10.1007/978-3-030-53587-2_1

themselves unchanging. Climate change is the death of stationarity (Milly et al. 2008). It's not just change; it is change in the way things change.

Stable ground is shifting like melting permafrost. The *permanent*, it turns out, isn't. The ship of civilization always rose and fell with the tides, but it was anchored to something deep. Now the bottom is falling out. We are falling. We are building such a heavy, such a weightless, world.

I wrote this book in dialogue with students in my college courses. One salient fact framed all of our conversations: Young people today are growing up on a different planet from the one I knew as a kid. A good way to see this is to look at the cumulative global emissions of carbon dioxide from fossil fuels. Since 1751, 1.54 trillion tons of CO_2 have been emitted. Note from Table 1.1 how long it took to emit the first quarter versus the last quarter.

CO_2 emissions reached record highs again in 2018 and 2019. This tells us two things. First, the human condition is accelerating. Second, we are not taking climate change seriously, which is to say that we are not reckoning with the speed or scale of our own actions. We *know* about the problem, but we don't really *believe* it. We have the science, but not the imagination. If ever there was a time to stop and think, well, now might be it.

That brings me to the second half of my thesis, which is about how this growing crack in reality *appears* to the denizens of a high-energy civilization. As energy grows bigger and stranger, things seem oh-so-normal. Like live wires wrapped in plastic, we are insulated from our powers. How easily we forget just how weird things are. We are yawning through a metaphysical revolution. After reading the dire headlines, we switch on the cartoons. It's so real it's unreal. So big yet so forgettable. Like I said, it's a paradox.

Climate change requires a change of mind. We have to live in the paradox, the *fullness* of our reality. This book aspires to help you do that.

Table 1.1 Cumulative global CO_2 emissions

Global CO_2 emissions	Historical period	Total years
First 25%	1751–1968	217
Second 25%	1968–1988	20
Third 25%	1988–2005	17
Fourth 25%	2008–2017	9

Source: Author's own table, data from Our World in Data, https://ourworldindata.org/co2-and-other-greenhouse-gas-emissions

It explores the origins of our high-energy civilization and the big questions it faces. My children were born on this new planet. When we discuss climate change as it appears, say, in California or Australia wildfires, they tell me it is scary. Then they ask, "Are we going to be ok?" That is the biggest question of them all.

To be sure, such questions have scientific, technical, and economic dimensions. Yet there is no formula to decide our future for us. We have choices to make, and they will hinge on our visions of moral responsibility, justice, freedom, knowledge, risk, and what it means to be human. The more powerful our science and technology become, the more philosophical issues they raise. Thinking through climate change is a philosophical task, one that requires us to dig down to fundamental issues and zoom out to see the contexts in which other ways of knowing (e.g., science, engineering, and economics) take shape (see Gardiner 2010; Gardiner et al. 2011; and Jamieson 2014).

There is so much information about climate change that it's like drinking from a firehose: overwhelming and confusing. I want to provide orientation by climbing up high, so that we can look down and see the many ways of seeing our situation. I categorize these ways of seeing or worldviews into the orthodox on one hand and the heterodox on the other hand. This is just a first-order divide, because there is diversity within both the orthodoxy and the heterodoxy. The orthodoxy deeply conditions how we think and act. That makes it worth understanding. However, the crack is growing and paradoxes are accumulating that might topple the orthodox order. That makes it worth considering heterodox views.

Here is the book in a nutshell. We are in a moment of exponential growth. Our future is either green growth or degrowth. Either we figure out how to make a project of infinite growth sustainable or we find some measure, that is, a sense of proportion and limit. The former is the orthodox view. The latter is the heterodox view. Energy consumption is expected to double by 2050. Clearly, we are gambling on the orthodoxy of green growth. Climate change is calling our bluff. We should understand the logic of the orthodoxy and pray that it is sound, because it is the hand we are playing in a game with existential stakes (Table 1.2).

Table 1.2 The book in a nutshell

	Discourse	Logic	View of humanity	Primary energy	Future	Ethics
Virtues	Heterodox	Doctrine of the mean, proportionality, limits	One among the earthly creatures	Control of self	Degrowth	Ends and means, fittingness
Volts	Orthodox	Infinity, linearity, growth	Gods in the making	Control of world	Green growth, decoupling	Means only, convenience

Now let me offer a more extended summary of the book. It begins with an obvious point that climate change is driven by energy. This is why most stories center on technology: fracking, solar panels, nuclear power, wind turbines, batteries, and more. (As we'll see, even the agricultural and land use dimensions of climate change are about energy. The conservationist Aldo Leopold (1945) was right to call the land "a fountain of energy.") The discussion is all about energy transitions, especially from fossil fuels to renewable or carbon-free sources. But in the debates about the means, we lose sight of the ends. In other words, this is all a debate *within* the orthodoxy, which is limited to instrumental ethics (i.e., we can evaluate means as better or worse, but not ends). To think through climate change, we have to understand energy in broader terms.

The most important energy transition is the one that took us from a world of virtues to a world of volts. Like any energy transition this is messy and incomplete, but it is vital. The virtues are intimately related to the original meaning of 'energy' in the West, one that denoted proportion or fit. The virtues are governed by the doctrine of the mean, which tells us when there is deficiency and when there is excess. There can be too little and *too much*. At some point, there is a phase change and, paradoxically, what was better is now worse. There is a limit, a threshold, a line you shouldn't cross.

I use 'volts' as shorthand for the modern scientific notion of energy. There is no upper bound to volts, no limit or sense of proportion. Its logic is linear, where things keep going up and up with no phase changes. The transition from virtues to volts, then, is from finitude to infinity. It brings with it a shift in our self-understanding from humans as one earthbound creature among others to humans as gods in the making. This is the metaphysical or religious story beneath the stories about energy and climate. The transition from virtues to volts is the golden thread that I trace in this book.

Our world of volts is the orthodoxy. We might also call it simply modernity or humanism. Here is the logic of the orthodoxy in a nutshell. Humans are weak in claw and muscle, but strong in brain. To survive, we figure out ways to control the Fates and their minions: cold, heat, hunger, disease, and aging. After millennia of searching, we have found the winning formula to set us on a path toward absolute security, control, and freedom. That formula is E for energy, the modern scientific notion of a universal currency and capacity to do work. (This is what I call 'volts' for short, it is the 'fire' that Prometheus stole, but this is the *real deal*.) We actually don't know what E is, but we know how it functions. We can measure it, quantify it, and exploit it to make our lives longer, healthier, more productive, more convenient, and above all more secure.

This is a story that begins in poverty and has its logical conclusion in the project known as transhumanism—the overcoming of all limits, including our bodies and our home planet. Because there is no upper bound or threshold to volts, growth is the grand totem. It's not just the essence of capitalism as a social order; it is the scientific picture of reality as a matrix of E and the ethical picture of progress as commanding more and more E. To get a sense of how strong the orthodoxy is, consider how crazy you'd have to be to run on a political platform of ramping *down* production and consumption. Yeah, right! The trajectory is "To infinity and beyond."

The titans of our economy and high priests of the energy orthodoxy know this. Bill Gates is pumping billions of dollars into research on endless, clean energy. Jeff Bezos, founder of Amazon and the world's wealthiest man, has said that his most important project isn't online shopping or streaming entertainment. Rather, it is Blue Origin, an aerospace manufacturing company that is making rockets for extraterrestrial resource extraction and space colonization. Bezos is worried that our growing energy demands will outrun our limited supply here on the third rock. Like me, he sees two basic choices: either we cap how much energy we use or we head for the stars. Bezos, the epitome of the energy orthodoxy, wants growth rather than stasis. His hero is Captain Picard from *Star Trek*—he has even shaped his appearance to look like Picard. As the tagline for Blue Origin reads, "Earth, in all its beauty, is just our starting place."

But what about climate change, that growing crack in reality? True, energy is about controlling fate and our scientific machines have given us so much control. Yet control is only half the picture, and as a result, the orthodoxy is a doctrine of half-truths. Powerful spells have a way of getting out of control. There are jokers in the high-tech hand we are playing.

The oldest stories in philosophy are about energy, and those stories are about a cosmos that is deeply ironic. The first philosophers tried to understand change: the seasons, the growing child, and the decaying fruit. They reasoned that there must be something that undergoes the change but is not itself changed. That is energy: an ever-changing sameness. A paradox. Energy is a wildcard; it's both the brute force of nature and her twisted sense of humor. Sure, any good book about energy and climate will have to be full of numbers. But it also has to account for what cannot be counted.

A high-energy society is bound to get tangled in its own contradictions. Paradoxes are springing up like the troubles from Pandora's Box. Before turning to the orthodox view, then, I start with some of the paradoxes that run like fissures through our bedrock certainties. In one of the first theories of energy, Heraclitus said that all is fire. Picture again the wildfires pulling civilization back to Earth and Bezos' rocket boosters heading for the stars. Which fire is our future? Trapped in indeterminacy like Schrödinger's cat, it's both.

Bibliography

Gardiner, Stephen. 2010. *A Perfect Moral Storm: The Ethical Tragedy of Climate Change*. Oxford: Oxford University Press.

Gardiner, Stephen, et al. 2011. *Climate Ethics: Essential Readings*. Oxford: Oxford University Press.

Heidegger, Martin. 1954. *Was heisst Denken? (What is Called Thinking?)*. Tübingen: Max Niemeyer Verlag.

Jamieson, Dale. 2014. *Reason in a Dark Time: Why the Struggle Against Climate Change Failed – And What It Means for Our Future*. Oxford: Oxford University Press.

Leopold, Aldo. 1945. *A Sand County Almanac, and Sketches Here and There*. New York: Oxford University Press.

Milly, P.C.D., et al. 2008. Stationarity Is Dead: Whither Water Management? *Science* 319 (5863): 573–574.

Part I

Energy Paradox

[I]n physics today we have no knowledge of what energy <u>is.</u>
Richard Feynman, Lecture on Physics, 1963

2

The Unnatural Growth of the Natural

We are inverted utopians: while utopians cannot produce what they imagine,
we cannot imagine what we produce.
Günther Anders *1956*

Human beings have become a dominant force on Earth. Many scientists believe that we have created a new geological epoch: the Anthropocene, or the age of humanity. Other scientists think that this is arrogance. After all, the title 'epoch' is given to thick stacks of rocks piled up across tens of millions of years. Yes, we are rearranging the face of the planet, but this is a mere blink of geological time. If we don't learn how to control the energies that we have unleashed, we may soon wipe ourselves out. In that case, all that we'll leave behind is a vanishingly thin line in the rocks. Geologists call such short-lived disruptions *events* not *epochs* (Brannen 2019).

Whether event or epoch, when did this new chapter in Earth history begin? Some think it started when hunters eradicated wooly mammoths and giant ground sloths. Others set the beginning at colonialism or the industrial revolution. One panel of scientists pegged it to the

© The Author(s) 2021
A. Briggle, *Thinking Through Climate Change*, Palgrave Studies in the Future of
Humanity and its Successors, https://doi.org/10.1007/978-3-030-53587-2_2

mid-twentieth century invention of nuclear weapons. Thousands of atomic explosions have carpeted the Earth with a telltale sheet of radiation. Alien archeologists in the future could visit here, dig down through layers of rock, and discover our signature written in plutonium-239. "Ah," they would say in their alien accents, "the Age of HUUMAHNS."

We are leaving other traces too, including micro-plastics and heavy metals. Industrial farming, deforestation, and massive dams alter landforms in ways that may leave a geological mark. The fossil record will show a precipitous drop in biodiversity, what many consider to be Earth's sixth mass extinction event (Kolbert 2014). Some few animals, however, will suddenly dominate the fossil record. The domestic chicken, for example, is native to south-east Asia, but in the Anthropocene their bones are piling up everywhere. We consume 60 billion chickens annually. The aliens might call this the Chickenocene.

Whatever we call it, no one can doubt the scale of human impacts or the speed with which they have happened. Indeed, some prefer to call this age the Great Acceleration. *Homo sapiens* has been on the planet for 200,000 years, but only in the last 200 years or so (0.1% of our history) have things gone crazy.

On graphs, the Great Acceleration looks like hockey sticks with their long shafts lying flat on the x-axis of time followed by the blade jutting steeply upward along the y-axis. The y-axis can represent socio-economic trends like human population, Gross Domestic Product (GDP) per capita, water use, fertilizer consumption, travel, and telecommunications. It can also represent Earth systems trends that show the same recent, sudden spikes: atmospheric concentrations of carbon dioxide and methane, ocean acidification, marine fish capture, surface temperatures, tropical forest loss, and species loss (Fig. 2.1).

Many factors are pushing those curves upward along the y-axis, but a central driver is energy. Our control of nuclear and fossil energies has fundamentally changed the human condition and our relationship to our home planet. This has happened very suddenly. For the vast majority of the human story, we had only the energy of our muscles, including the muscles of slaves. About 10,000 years ago we harnessed the energy of animal muscles. Over time we invented waterwheels and windmills.

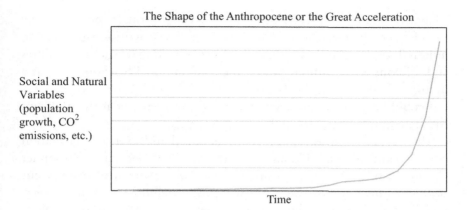

Fig. 2.1 The shape of the Anthropocene or the Great Acceleration

These energy transitions introduced big changes. The energy analyst Vaclav Smil (2010) estimates that peak unit capacities of prime movers rose by a factor of 150,000 in the 3000 years prior to the twentieth century. That's impressive. But those 3000 years pale in comparison to just the last one hundred. In the twentieth century alone, peak capacities rose ten times as much, by a factor of 150,000,000! Like the Sorcerer's Apprentice, we have cast a powerful spell that threatens to get out of control. In that story, the master returns in the nick of time to save the day. We, by contrast, are on our own.

We consume 100 million barrels of oil globally every day. And global energy consumption is expected to double by 2050. In the United States, natural land is being converted into human development at a rate of two football fields every minute (Lee-Ashley 2019). Roughly 70% of the Earth's surface has been shaped by human activities. Urbanization or the building of the 'technosphere' is proceeding at a breakneck speed. We are going to build the equivalent of a new New York City *every month* for the next thirty years. China poured more cement from 2011 to 2013 than the United States did during the entire twentieth century (Smil 2013). Roughly 8 million tons of plastic are washed into the ocean annually, meaning that by 2050 plastics might outweigh all the fish in the ocean (World Economic Forum 2016). Biologists on the remote Midway

Islands estimate that every year albatrosses carry 5 tons of plastics to the islands *in their stomachs* (Alfonsi 2019). As I was writing this chapter, hundreds of thousands of Californians were forced to flee hellish wildfires and millions went without power due to intentional blackouts. It was a dystopian scene. The new abnormal.

I could keep listing the stats, but that's enough to get at the problem. All these numbers are so big that they defy belief.

The scale of the Anthropocene and the speed of the Great Acceleration pose a fundamental dilemma spotted by the German philosopher Günther Anders early in the atomic age. For nearly all of history, our abilities to imagine (*vorstellen*) outstripped our abilities to produce (*herstellen*). We could dream big, but we lacked the energy to build big. Now, things are inverted. Our productive powers exceed our imaginative ones. We are making a world that we cannot comprehend. The scholar Timothy Morton (2013) puts this in terms of 'hyperobjects,' phenomena that are so massively distributed across time and space as to confound our usual way of making sense of things.

Climate change is the prime example. It is there in the flood or the wildfire, but it is also not there. We can neither escape it nor keep our attention trained on it. Despite billions of dollars of scientific research, we have still never *experienced* or *felt* the climate. What we experience is weather, and it's always changing, so what's the big deal? That might explain why fossil fuels remain at around 80% of the world's energy mix—the same as it was back in 1987 (Harder 2019). The global economy hasn't decarbonized any faster during the era of climate science than it did in the two decades prior to all that knowledge (Pielke 2019).

Are we even *capable* of grasping what we are doing? As Nietzsche asked in his parable of the madman from *The Gay Science*, "Is not the greatness of this deed too great for us?" (1882, para. 125).

Climate change is everywhere and nowhere. It is now, but it can't be now because the *now* is the time of weather. After Hurricane Dorian devastated the Bahamas in 2019, the homeless survivors looked like victims of bad weather rather than climate. You can see how we might react to a climate apocalypse like the proverbial frog in the boiling pot of water comfortably slipping into oblivion.

Before he was forced to flee Nazi Germany in 1933, Anders married the political thinker (and fellow student of Martin Heidegger) Hannah Arendt. In her 1958 book *The Human Condition*, Arendt worried that we may soon no longer be able to "understand, that is to think and speak about the things which nevertheless we are able to do."

Hans Jonas (1984), another student of Heidegger's and a lifetime friend of Arendt, argued that all previous ethics could assume "that the range of human action and therefore responsibility was narrowly circumscribed." Our high-energy machines have altered the scale of our action and since ethics has to do with action, our ethics must change. But we may simply not be wired for this. If you strap someone into a functional magnetic resonance imaging machine (fMRI) and watch as they think about themselves, their medial prefrontal cortex lights up. We *care* deeply about ourselves. The lights get dimmer and dimmer as we think about people further removed from this central ego—family, friends, and acquaintances (see Walsh 2019). Thinking about a stranger in the Bahamas who lost their home hardly creates any spark at all.

It's not just spatial scales that challenge our moral psychology. It's also time. The prefrontal cortex even dims when you think about yourself in the future. As economists know, we discount the near future, which means it is worth less. The far future is entirely worthless, but of course what we call the "far future" is no time at all for the planet. There's the problem: we are geological agents unable to think geologically. Time and space are slipping from our grasp. This is why "global weirding" is a good term for what is happening.

Anders (1957) wrote that your first thought upon waking up in the morning should be 'Atom.' You should call to mind the enormous powers pulsing under the seemingly steady day-to-day world. "For you should not begin your day," he continued, "with the illusion that what surrounds you is a stable world." Your second thought should be: "The possibility of the Apocalypse is our work. But we know not what we are doing." Even the experts are ignorant when it comes to the whole. We cannot "realize the reality which we can bring into being." There is a gap between our actions and our imagination. Weird things are falling through the crack.

What was a gap in the time of Arendt and Anders is now a chasm. We have altered the energy balance of the entire planet. Now our first thought on waking in the morning should be CARBON, because the extra energy trapped in the ribbon-thin atmosphere from greenhouse gas emissions is equivalent to that released by 400,000 nuclear bombs exploding *every day* (Hansen 2012). We are proving that Heraclitus was right with his original philosophy of energy: all is fire.

* * *

The challenge facing us is unlike any other in human history: We have to fathom the world that we are making. The understanding that we require is not just scientific or technological, and the 'energy' we have to fathom is not just the stuff stored in chemical bonds, flowing through pipelines, or buzzing across wires. The energy transition that caused the Anthropocene is not just the one to nuclear power and fossil fuels. Something deeper happened over the last few hundred years. There was a revaluation of values, a transition from a world of virtues to a world of volts. As a result, the metabolic energies of human labor and consumption became unhinged and we started eating the planet.

Living organisms constantly wrest themselves from non-self. They pull in air and nourishment for their metabolic fires. The basic energies of life are labor and consumption (which are two stages of the same process), and they are incessant. I will feel the pangs of hunger tonight even though I just ate lunch. It will come again in the morning and for the rest of my life. The caloric demands of the consuming and laboring body cease only upon death. But this incessant activity, Arendt argued, is one that spins round and round a wheel and in that sense it "remained stationary" for all of human history. All along the long shaft of the hockey stick, human energy was "imprisoned in the eternal recurrence of the life process to which it was tied" (Arendt 1958, p. 46).

The jutting blade of the hockey stick, the Great Acceleration, is the rupturing of the chains that had kept this incessant life process spinning cyclically. Labor and consumption were taken from the dark interior of the private realm and admitted into the public realm. The household economy became the global economy. Labor, Arendt notes, has become liberated "from its circular, monotonous recurrence and transformed...into a swiftly progressing development whose results have in a few centuries totally changed the whole inhabited world" (p. 47). As Nietzsche put it, we lost all sense of limit or measure and embraced "the thrill of the infinite." We went from a world of virtues to a world of volts.

Arendt thought that this was a paradox. She called it "the unnatural growth of the natural," because the natural metabolic energy transformations were unleashed beyond all natural boundaries. We call this economic growth. Just as I need to feed my own metabolic fire with lunch, we need to feed this global metabolic fire. The difference is, though, that my body only grows to a point and then stops. The collective body of the Anthropocene keeps growing beyond all proportion and measure. There is no limit, no sense of sufficiency.

Can this continue? As I was writing this book, climate change morphed into the climate crisis. Greenland experienced a record ice melt. There were record heat waves in Europe, and unprecedented fires and floods around the world. The hockey stick blades kept reaching alarming new highs. Despite climate treaties, global CO_2 emissions kept going up. The Intergovernmental Panel on Climate Change (IPCC), the world's most respected source of climate science, issued new studies with grim news about the impact of the human economy on planet Earth. Saying that they have a "moral obligation" to "tell it like it is," over 11,000 scientists broke with the more conservative rhetoric of the IPCC and issued a statement proclaiming that "Earth is facing a climate emergency" (Ripple et al. 2019). Prominent voices warned that the Earth may soon be uninhabitable (Wallace-Wells 2019) and that the human game may be playing itself out (McKibben 2019).

If this is an emergency, then it would seem that this unnatural growth of the natural must end. We must impose limits and pull on the reigns. Goodbye capitalism, materialism, and consumerism. Ration hamburgers and airline travel. Good night, Disney World. Adios, summer vacation. Smil (2019) puts it plainly: growth must end. It is time for another revaluation of values. A new way of life.

Yet with the exception of a few heterodox voices (e.g., Meadows et al. 2004; Kallis 2011), no one is proposing an agenda of limits. The American Green New Deal, for example, is a sweeping proposal to transform the energy sector to achieve carbon neutrality. It calls, among other things, for a massive redistribution of wealth. It has been criticized for being too radical and socialist. Yet it is actually the same old stuff of regulated techno-capitalist enterprise. It is still a manifesto for laborers and consumers; it's just *green* labor and consumption. When the critics said that the Green New Deal would force people to give up meat, its defenders were quick to insist that it did no such thing. You can still have your cake and eat it too. Just switch to renewable energy, then go back to your online shopping. Same lifestyle, just carbon neutral.

The economist Ted Nordhaus (2019) writes that the so-called radical environmentalists insist "that capitalism and technology are the problem, not the solution to our present predicament when practically, after the sloganeering and rhetorical flourishes are done, what most environmentalists, including radical greens, are basically demanding is capitalism with carbon regulations and lots of windmills." Amory Lovins and Rushad Nanvatty (2019) similarly argue that "any serious energy transformation effort" needs to "harness America's immensely powerful and creative economic engine, not dismantle it." That engine may have brought us to the brink of catastrophe, but it's also the only thing that can save us now.

If we look past alarmism to see how people act and what they actually propose doing, we discover a deep consensus. Nordhaus puts it this way: "Practically, we are all neoliberals now. Some of us just haven't realized it." Almost all of our political debates happen within a shared framework that says, roughly, that the problems caused by innovation and growth can only be solved by further innovation and growth. More jobs building more wind turbines. More investment for more nuclear, concentrated solar power, and grid-scale battery storage. Even the most famous formulation of 'sustainability,' the 1987 Brundtland Commission Report, is clear that we can and should "make way for a new era of economic growth." We need to decarbonize the economy, not degrow it.

I call this shared framework the "energy orthodoxy," because it is so widely held and it seems to strike most people as obviously right (orthodox means right belief). Above, Anders wrote that we are the creators of the apocalypse and we know not what we are doing. The orthodoxy tells us to delete just one word. Take out the 'not.' Yes, we have conjured destructive powers. But we *control* them. That is, we *know what we are doing.* Is this *hubris* or is it just the calm, puzzle-solving voice of science? Do we know what we are doing? Our fate hangs on that question.

To help us fathom the world we are making, I put the orthodoxy to the test. This first section, "energy paradox," prompts us to think outside of the orthodoxy. Then, I switch voices in the middle section to defend the orthodoxy. The last section, "energy heterodox," surveys ways to think past our usual way of thinking. If this is the age of climate weirding, then we may be in for a weird new way of life. It's good to think some strange thoughts.

*   *   *

Arendt said that modern civilization represents the rise of humans as the *animal laborans*, because labor and consumption are our highest ideals. We might update that now to *Homo* ludens (from the Latin for 'play'), given how entertainment has grown so much since the mid-twentieth century when she was writing her cultural criticism. Neil Postman (1985) is right about "amusing ourselves to death." The paradox is how a high-tech culture can be so low-brow. Tech giants and moral midgets.

The first section of this book is all about paradox, inspired by the original mystery of energy as that which changes yet remains the same. The trouble with modern energy has always been the way it conceals nature's crooked smile. In his history of energy, Crosbie Smith (1998) shows how our modern orthodox work-based notion of energy allowed for a system of measurement that served scientific, imperial, and commercial interests. The old language of 'forces' was too metaphysical and couldn't allow engineers to translate the science into machines. Modern science gave up on the quest to know what energy *is* to settle for knowing what it *does* (see Rowland 2019).

This has enabled the Anthropocene and the Great Acceleration. But it is also obviously a kind of fudge. It is a mask that shows a very serious and stern face of quantification, standardization, and control. Energy, though, is two-faced. There is another mask, that ironic smile of the mysterious 'force.' Being two-things-in-one is the nature of paradox. Table 2.1 collects the key paradoxes that we'll explore.

Table 2.1 Some paradoxes of a high-energy civilization

Paradox	Notes
Unnatural growth of the natural	Labor or metabolism is turned inside out. Our lowest form of energy, labor, becomes our highest ideal
Strength-weakness	Strong machines, weak wills. Controlled nature, uncontrolled desires. The non-negotiable life of negotiations
Thick world, thin places	The not-here is here. The not-now is now. The walrus you shoved and never touched
Unbelievable knowledge	Knowing more and more; understanding less and less. Unfathomable world of our own making
Active passivity	The more that is done, the less anyone is responsible for doing. The great no one of collective (in)action. The innocent guilty
Enlightenment shadows	Rationality breeding irrationality; precise understanding and control of reality breeding multiple realities; denying reality and embracing conspiracy
Collective individualization	The ego is insulated and isolated even as it is stitched and tethered ever more tightly to the collective. Being alone together in the new commons
Poverty of the wealthy or the modern poor	Worm energy and divine energy as inversely related; the good life as a treadmill, the forever horizon. The black hole of needs. We become the tools of our tools
The immorality of justice	Justice is chained to growth, her scales tipping out of balance. We must help the poor to participate in self-annihilation just like the rich

Paradox means not just a statement that is contrary to common belief, but also a statement that seems self-contradictory but is not illogical or obviously untrue. The physicist Richard Feynman (1963) said of energy, "it is just a strange fact that we can calculate some number and when we finish watching nature go through her tricks and calculate the number again, it is the same." Heraclitus insisted that the river is the same *and* changing. He put the paradox this way: "Everything always has its opposite within itself."

This poses a metaphysical dilemma. Something can't be A and not-A, so just what *is* it? In politics, this metaphysical question becomes a matter of *framing*. For example, surveys during the 1980s found that 60% of Americans agreed that the federal government spent too little on "assistance to the poor." Yet only about 20% said that the government spent too little on 'welfare.' Someone might favor the "estate tax" but not the "death tax." Of course the classic case is abortion. Just what *is* it: a choice or a life? When a baker refuses to make a cake for a same-sex couple, is that discrimination or religious liberty? Paradox is the stuff of *both-and*.

In her own book about paradoxes, Deborah Stone (1998) argues that politics is a battle of ideas as factions attempt to resolve paradoxes one way or the other. Political reasoning is reasoning by analogy and metaphor. It is the art of getting people to see a situation *as* one thing *rather than* another. Imagine that we have been hired to measure an elephant. There are many possible measures, including height, weight, volume, trunk length, intensity and range of skin colors, depth and number of wrinkles, time spent sleeping, intestinal surface area, and so on. Imagine all the different numbers that could be produced. All of them would depend on prior judgments made in answer to the question: Just what *is* this elephant that we are measuring? It is like the blind men who insist it is a fan, a snake, a tree, a rope, or a wall, depending on what part they are touching.

Stone's point is that some selection must first be made about what to count before the counting can happen. Numeracy is doubtlessly crucial for the task of fathoming our world. Like I said, any good book on energy will be full of numbers and units. But numbers are the masks that values wear, so we have to understand not just the numbers but the judgments behind them. Why these numbers? Which numbers are relevant? What are we choosing to measure? Whose interests does that serve? The point is *not* that all framings are equally true. Indeed, we must resist false equivalencies. A paradox is a genuine predicament and cannot be created by simply asserting "not-A" in the face of overwhelming evidence for 'A'. Just saying "climate change is a hoax" doesn't create a paradox; it creates disinformation. The point, rather, is that getting to the truth is not as simple as letting the numbers or the science speak. Which numbers, whose science?

This is why philosophy and the humanities are fundamental to our political debates about energy and climate (Szeman and Boyer 2017). Issues are always framed this way or that way, which highlights some values and downplays others. This means that as we exercise our moral imagination to fathom our world, we need to not just see the Anthropocene but see *the many ways of seeing* (or framing) the Anthropocene. We need to see, not just energy, but the many faces of energy. That's why I focus on *doxa*. It doesn't just mean 'opinion,' it means "the way things appear." Look at the victims of Hurricane Dorian again—do you see climate refugees? *Who are they?*

Once a frame has been chosen, the question of what something *is* is no longer asked. The paradox is resolved. Metaphysics is set aside. Judgments and values are buried under numbers. Certainties set in. Deeper questions about ends are black-boxed to focus on the means. A canon of acceptable language, topics, and metrics is adopted. This is how orthodoxies like the one about energy are established. And as Heidegger would say, this is how *Denken* (thinking) is replaced by calculation. My goal with this book is to keep the paradoxes alive, thus the emphasis on *thinking* through climate change.

Maybe the orthodoxy is right that we can tech-fix and grow our way into an enduring human epoch. Maybe we really have no other choice. Then again, maybe we just can't imagine how deeply we have deranged the order of things. The assumptions buried in our metrics might be rising like zombies from the dead. The one metric we lack is the one that Nietzsche called 'measure' as in *proportion*. It is the one that draws a line marked: ENOUGH. Maybe this line will be revealed, like a divine truth, under the receding glaciers. Yet even if we saw it there etched in stone, would we *believe* it? Even if we quantified it, would it have the qualities needed to *move* us?

In quantum physics, a superposition happens when a particle exists in all possible states at the same time. It is A *and* not-A. The US Department of Energy is jumping into quantum science now. They are afraid that China is getting the edge. The goal is to go beyond the digital age into the quantum age. After bits, we will master qubits, which behave like a spinning penny: heads *and* tails, zero *and* one. The future of high-energy civilization is paradox.

The Enlightenment casts shadows. High-powered machines breed powerlessness. Energy slaves become masters. The Anthropocene, this age of humans, puts us at the center of things only to demonstrate our own futility. We are like the great Prometheus who stole fire from the gods. We are on prominent display, to be sure, but only to witness our own liver being devoured.

Bibliography

Alfonsi, Sharyn. 2019. Cleaning up the Plastic in the Ocean, *60 Minutes*, August 18. https://www.cbsnews.com/news/the-great-pacific-garbage-patch-cleaning-up-the-plastic-in-the-ocean-60-minutes-2019-08-18/

Anders, Günther. 1956. *Die Antiquiertheit des Menschen. (The Outdatedness of Human Beings)*. Munich: Beck.

———. 1957. Commandments in the Atomic Age. In *Burning Conscience*, 11–20. New York: Monthly Review Press.

Arendt, Hannah. 1958. *The Human Condition*. Chicago: University of Chicago Press.

Brannen, Peter. 2019. What Made Me Reconsider the Anthropocene. *The Atlantic*, October 11. https://www.theatlantic.com/science/archive/2019/10/anthropocene-epoch-after-all/599863/

Brundtland Commission. 1987. *Our Common Future*. https://sustainabledevelopment.un.org/content/documents/5987our-common-future.pdf

Feynman, Richard, Robert Leighton, and Matthew Sands. 1963. *The Feynman Lectures on Physics*. Vol. 1. New York: Basic Books.

Hansen, James. 2012. "Why I Must Speak out on Climate Change," *Ted*, February 29, transcript at https://blog.ted.com/why-i-must-speak-out-on-climate-change-james-hansen-at-ted2012/.

Harder, Amy. 2019. Why Climate Change is so Hard to Tackle: Our Stubborn Energy System. *Axios*, August 26. https://www.axios.com/why-climate-change-is-so-hard-to-tackle-our-stubborn-energy-system-6c8fc596-3c47-477a-82aa-cd00f063c9a0.html

Jonas, Hans. 1984. *The Imperative of Responsibility: In Search of an Ethics for the Technological Age*. Chicago: University of Chicago Press.

Kallis, Giorgos. 2011. In Defence of Degrowth. *Ecological Economics* 70: 873–880.

Kolbert, Elizabeth. 2014. *The Sixth Extinction: An Unnatural History*. New York: Henry Holt and Co.

Lee-Ashley, Matt. 2019. How Much Nature Should America Keep? *Center for American Progress*, August 6. https://www.americanprogress.org/issues/green/reports/2019/08/06/473242/much-nature-america-keep/

Lovins, Amory, and Rushad Nanvatty. 2019. A Market-Driven Green New Deal? We'd Be Unstoppable. *New York Times*, April 18. https://www.nytimes.com/2019/04/18/opinion/green-new-deal-climate.html

McKibben, Bill. 2019. *Falter: Has the Human Game Begun to Play Itself Out?* New York: Henry Holt and Co.

Meadows, Donella, et al. 2004. *Limits to Growth: The 30 Year Update.* London: Earthscan.

Morton, Timothy. 2013. *Hyperobjects: Philosophy and Ecology after the End of the World.* Minneapolis: University of Minnesota Press.

Nietzsche, Friedrich. 1882 (1974). *The Gay Science.* Trans. Walter Kaufmann. New York: Penguin.

Nordhaus, Ted. 2019. The Empty Radicalism of the Climate Apocalypse. *Issues in Science and Technology*, vol. xxxv, no. 4. https://issues.org/the-empty-radicalism-of-the-climate-apocalypse/

Nordhaus, Ted, and Alex Trembath. 2019. Is Climate Change like Diabetes or an Asteroid? *The Breakthrough Institute*, March 4. https://thebreakthrough.org/articles/is-climate-change-like-diabetes

Pielke, Roger. 2019. The Yawning Gap between Climate Rhetoric and Climate Action. *Forbes*, September 19. https://www.forbes.com/sites/roger-pielke/2019/09/19/the-yawning-gap-between-climate-rhetoric-and-climate-action/#2e8de4ed2ec4

Postman, Neil. 1985. *Amusing Ourselves to Death: Public Discourse in the Age of Show Business.* New York: Penguin.

Ripple, William, et al. 2019. World Scientists' Warning of a Climate Emergency. *Bioscience*: biz088. https://doi.org/10.1093/biosci/biz088.

Rowland, Jennifer. 2019. Conceptual Barriers to Decarbonization in US Energy Policy. Unpublished dissertation, University of North Texas.

Smil, Vaclav. 2010. Science, Energy, Ethics, and Civilization. In *Visions of Discovery: New Light on Physics, Cosmology, and Consciousness*, ed. R.Y. Chiao, M.L. Cohen, A.J. Leggett, and C.L. Harper Jr., 709–729. Cambridge, MA: Cambridge University Press.

———. 2013. *Making the Modern World. Materials & Dematerialization.* Chichester: Wiley.

———. 2019. *Growth: From Microorganisms to Megacities.* Cambridge, MA: MIT Press.

Smith, Crosbie. 1998. *The Science of Energy: A Cultural History of Energy Physics in Victorian Britain.* Chicago: Chicago University Press.

Stone, Deborah. 1998. *Policy Paradox and Political Reason.* New York: Scott Foresman & Co.

Szeman, Imre, and Dominic Boyer, eds. 2017. *Energy Humanities: An Anthology.* Baltimore: John Hopkins University Press.

Wallace-Wells, David. 2019. *The Uninhabitable Earth: Life after Warming*. New York: Tim Duggan Books.

Walsh, Bryan. 2019. Why Your Brain Can't Process Climate Change. *Time*, August 14. https://time.com/5651393/why-your-brain-cant-process-climate-change/

World Economic Forum. 2016. *The New Plastics Economy: Rethinking the Future of Plastics*, January. http://www3.weforum.org/docs/WEF_The_New_Plastics_Economy.pdf

3

Walrus Guts and Snake Brains

The American way of life is not up for negotiations
George H.W. Bush at the 1992 Earth Summit in Rio de Janeiro

It was getting hot, so I pushed the down-arrow on the thermostat for the air conditioner. A snowflake icon appeared on the screen just before the walrus crashed through the roof. Its blubbered brown body landed in the living room with a thunderous boom. It burst into smithereens. Walrus guts hung from the television and tusks lodged in the leather couch. Along with a disembodied walrus eye, I gazed up through the hole in the roof, wiped some blood off my computer, and got back to writing the book.

A little while later, I heard a commotion: on my driveway outside the window a woman appeared from the tailpipe of my white minivan like a genie from a bottle. Her sari swayed as she walked into the house. In one hand, she held a small child. In the other hand, she held a cardboard sign that read: "Climate refugee. Will work for *terra firma*." She calmly picked her way through the smashed walrus and helped herself to the guest

A. Briggle, *Thinking Through Climate Change*, Palgrave Studies in the Future of Humanity and its Successors, https://doi.org/10.1007/978-3-030-53587-2_3

room. Once she was settled in with a nice glass of wine, she opened her suitcase and a river poured out in a flood.

I floated on my desk to the fridge to grab a snack. Upon opening the door, the genome of the soybean pounced out. It moved like an abstract shark and gnashed at me with teeth made of patents. So cute! I patted its head and turned back to the computer to write the book. By then, a drought had taken hold and the walrus' body had desiccated. Robots helpfully picked up the bones so that I could focus on my academic work. Back to the book! My desk was on fire; I typed on melted keys.

The television blinked on. I couldn't help but stare listlessly at the screen. Just another ho-hum day of news. A bicycle rider in the *Tour de France* ripped his cold suit, allowing the blistering Parisian air into his lungs. He collapsed on the Avenue des Champs-Élysées while other riders whizzed by, celebrating the end of the race with a steaming hot cup of champagne. Meanwhile, a hail storm in the south of France had shut down parts of the race course with piles of ice. "They don't call it global weirding for nothing!" chuckled the commentator.

In other news, a squid was refusing to pay rent for his new underwater apartment in Miami. The cause of the complaint? "There are barnacles clogging up the toilet." It had become a real problem. Elsewhere, the elk in the District-Formerly-Known-As-Glacier-National-Park were on strike. They refused to bugle, prance, or clack their antlers until their bottled water rations were increased. Meanwhile, tourists were entertained by new mechanical elk, which according to one woman from Kansas, "are even better than the real thing!" They produced no unseemly smells.

Agh, the book! Focus, focus. Yes, it was a book about climate change and energy. I poised my fingers to type, but then a voice from the television grabbed hold of my attention. We were back in Paris. That day, it was $109°$ F—a mere $32°$ above average—as if *average* had any meaning left. The voice on the TV belonged to a young urban planning consultant who had abandoned her sweltering office. An American news crew caught up to her and asked if she was going to buy an air conditioner. After all, 90% of American households have an air conditioning system but only 10% of European homes do.

The young woman said 'no.' The heat wave, she suggested, offered a lesson: "It's better not to have air-conditioning so we can feel the real effects of climate change" (Magra et al. 2019). Her words haunted me. They kept me writing the book when, perhaps, I should have been at a protest or installing solar panels, or *doing* something. "Feel the real."

Energy allows us to *control reality*—to turn hot air into comfortable cool air. But this enables us to *ignore reality*—to hide from the consequences of our controlling actions. The problem is that we won't feel— we won't *realize*—when our control is getting out of control. We won't recognize how reality is getting weird, even *unreal*. We won't believe it. We won't fathom it. That young woman rejected the air conditioner as a magic trick; now is the moment to say 'no' to illusions. Now it is time for reality. Not so much to know it, but to *feel it*.

Refuse the energy of air conditioners and embrace the energy of compassion and citizenship-as-fellow-Earthlings. Was she telling me the thesis of my book? I took notes. Scratched my head with a gleaming branch of bleached coral reef.

Is the walrus real? Do you feel the thud of its cantankerous poundage hitting the fake wood floor? Think of this in terms of "thin places." These are special places where heaven and Earth or hell and Earth cross wires. They might be a sacred building or a holy site in the forest or on a mountaintop. They are spots where different energies intermingle. Our side, the realm of daily life, is the right-side-up. The other side, the mystical, is the upside-down. Thin places are portals linking our everyday reality to the realm of the weird and stranger things.

For most cultures, these places are rare and require careful attendance. But our globalizing high-energy culture creates millions of thin places. The upside-down behind the thermostat, tailpipe, and fridge is a monstrous world of wires and fossil fuels. The push of a button on this side activates armies of hot-breathed machines on the other side. The walruses are famous actors from the Netflix documentary *Our Planet*. In desperation, they crowd onto a small cliff in the arctic because those machines have melted the sea ice where they once spread out. They jostle at the edge, lose their footing, and plummet to their deaths. The cameraman falters slightly in recording the last weightless seconds of one massive creature after another.

Who pushed them? Not you or me. No one did. The Great No One.

The more we channel powerful upside-down energies, the less we believe in them. Those energies build a *world* so seemingly 'thick' that it leaves us doubting whether this is really *our planet* at all. Walruses may fall, but always *somewhere else*. But that's the illusion. That young woman on the television was saying that we need to listen. Because *somewhere else* has just come through the portal and landed *here*. This is our unfathomable quantum superposition. The walrus is here *and* not here. What we need are Earthlings who can handle that.

*   *   *

When President Bush said that our way of life is not up for negotiation, he was naturally voicing American arrogance. But he was saying more than that. After all, the "way of life" at issue is the ethos of far more than one nation. It is modernity—what we mean by progress and development. The Chinese and Indians are building air conditioners like crazy. Their homes, like mine in Texas, also get hot. Why should they not be entitled to their own little snowflake icons? The high-energy way of life is good and everyone deserves it. Everyone *needs* it. Justice demands it. *That* is not up for negotiation.

This non-negotiable way of life—this *ethos*—is stamping its image across the planet. It mines, fracks, nukes, drills, covers the desert in solar panels, and erects wind turbines that make the Statue of Liberty look like a little baby doll. It casts mushroom shadows on the white sands of Alamogordo, scratches oozing tar scabs into the boreal forest, and paints the sheen of gasoline rainbows on the Niger Delta. It stuffs the Louvre and the Vatican and Yellowstone overfull with tourists fresh off of jet planes. And, of course, it is weirding the climate.

I have been calling this ethos the energy orthodoxy, because it is fueled by faith. Its faith is not invested in a deity, though. It is a faith in ourselves—that ours is a righteous cause and that we are clever enough to master ever-increasing flows of energy. This is faith in science and technology. The walruses are a test for the faithful.

How we act depends on what we think and believe, or our *doxa*. If ethics is an investigation of how we should act, then it begins with questions about what we should believe. Try reading the comments from an online story about climate change. Notice how people often radically disagree with each other about the state of the planet. If we all live in the same reality, how can it appear so differently to people? What do you believe in? Magic, voodoo, quarks, the Department of Energy, *The New York Times*? Is Greenland really melting? How would you know—have you seen it with your own eyes?! Fake news!

The word 'orthodox' means right-belief or right-opinion. It is the position largely held to be correct, that is, simply an accurate reflection of reality. Another name for the orthodoxy is the Overton Window: the boundaries that distinguish legitimate discourse from nonsense. It tells us what is a serious plan and what is a foolish daydream.

We know there is an energy orthodoxy, because some things are treated with utter seriousness by people in important agencies wearing nice suits while other things are dismissed as whacky. The orthodoxy is why we guffaw at proposals for slowing down or shrinking the economy yet we don't even titter when scientists talk about spraying sulfates into the stratosphere to cool down the planet. Indeed, we throw money at them! Limits are laughable, lunar mining excavations for Helium-3 are laudable.

Philosophers throw rocks at Overton Windows. Our job is to see if people can or should think in new ways. Arguably, our high-energy age with its cascading global environmental and social effects is so disruptive that it calls for us to think and believe and live differently. Faith in technoscience is not utterly misplaced, but perhaps it should become less and less central, because it is getting dogmatic.

A case in point: many climate models assume that carbon capture and storage technologies will work economically at a massive scale quickly. That could happen, but it strikes many analysts as a total pipedream (e.g., Heinberg 2018). We are in trouble when our most serious scientific reports come wrapped around a core of wishful thinking. Continued economic growth is also taken dogmatically as an unquestioned premise, but some analysts (e.g., Smil 2019) doubt that we can long sustain that illusion. We are putting all our chips on a radical technological breakthrough and status quo growth. We may want a back-up plan.

It is getting hotter and weirder. In response, yes, we can hit the down-arrow on the air conditioner. But we can and should also open the Overton Window a bit wider. Our times call for some fresh metaphysical air, some new beliefs. In other words, we can seek tech-fixes *and* spiritual fixes. We can say 'no' to air conditioners in order to "feel the real." We can laugh a little at the orthodoxy and take paradoxes and heterodoxies more seriously.

But wait. Everyone needs more energy. That is the crusade, the moral imperative, of the orthodoxy, and it is no laughing matter. We need this energy not just for life but for the *good* life. Now, you won't hear much talk in the orthodoxy about what constitutes the good life. That's because it is a matter of dogma. We all *know* what happiness entails. The ends are obvious, the only question is how to secure sufficient means. The orthodoxy deals exclusively in instrumental rationality: *given* the ends, what tools do we need to achieve them? The paradox is in how we become the tools of the tools.

Like any orthodox faith, the energy orthodoxy has its high churches where official doctrines are created and disseminated. One of the main hubs for the energy orthodoxy is CERAWeek, the world's leading energy conference. This annual event held in Houston brings together the most respected energy gurus from around the world—oil ministers from Saudi Arabia, energy czars from Russia, politicians from Azerbaijan, and so on. At the conference, you will be treated to a wealth of expertise about the latest technologies, the supply chains, the regulatory structures, and so on.

Much the same holds true for the other high-churches of the orthodoxy. The US Energy Information Administration, for example, offers reams of data on sources and uses—where the energy comes from (petroleum, nuclear, solar, etc.) and where it is used (transportation, industry, electricity, etc.). The International Energy Agency offers all manner of statistics and metrics. Production yields and forecasts. Consumption outlooks. BTUs, CO_2, kilowatts, short tons, barrels. This is the orthodox language of energy. And it is the orthodox language of the *ethics of* energy, because more energy is better. Energy literally empowers the good life.

This language, however, speaks to just one *kind* of energy. It is the kind that can be tracked and measured. There are other energies that matter. In the midst of his own calculations in *Walden*, Henry David Thoreau noted how nations can become obsessed with moving and hammering stones to memorialize themselves. "What," he asks, "if equal pains were taken to smooth and polish their manners? One piece of good sense would be more memorable than a monument as high as the moon."

Of course, good sense and compassion are harder to measure than monuments or megawatts. They don't fit in the Overton Window. Yet, if we think a little wider, we can read Thoreau's *Walden* as a philosophy of energy. He begins with a reflection on "vital heat" and the necessities of food, clothing, and shelter that ensure a supply of that life energy. Then he asks a curious question: what about when we have *enough* of those necessities? Surely, then we shouldn't busy ourselves with "more warmth of the same kind, as more and richer food, larger and more splendid houses, finer and more abundant clothing...and the like." No, we should attend to a higher kind of life:

> When he has obtained those things which are necessary to life, there is another alternative than to obtain the superfluities; and that is, to adventure on life now, his vacation from humbler toil having commenced.

The energy orthodoxy has given millions of people this "vacation from humbler toil." They lead lives of comfort and ease unimaginable to people in the past. Yet the orthodoxy has lost all sense of this other kind of energy and this higher kind of life. It can offer only further instruments for pursuing ever more of the necessities of life—bigger houses, more clothes, and something Thoreau could not have imagined: endless entertainment. In this way, the orthodoxy slides seamlessly into the 'superfluities.' We come to *need* more and more. The rational pursuit of the means of survival undergoes a phase change into the irrational elaboration of needs. What was once excessive becomes essential to keeping the economy humming along.

We have marvelous time-saving and labor-saving devices but we are busier than ever before. We are not sure what all this time is being saved *for*. The orthodoxy tells us about which resource to exploit and how to do it, but far less about why or wherefore. You can learn how fast we are going, but not whither. Everyone knows *where* we should go, right? Onward and upward! But we have forgotten which way is up.

The energy orthodoxy, like all dogmas, insists that it knows. It knows what the good life is: comfort, material prosperity, and convenience. But is that right? It knows what energy is: simply "the capacity to do work." But it fails to see the full depths and dimensions of this work. It is all hubris, no humility. All metrics and no measure. An algorithm without a rhythm. It holds a lock on our imaginations, but the thud of falling walruses (the other internet-of-things) is getting closer. The fires are in Malibu. Miami is sinking under the party. It is time to either think as big as we act or start acting as small as we seem capable of thinking.

*   *   *

The orthodoxy, our way of life, is itself a paradox. It is simultaneously about strength and weakness. It is the result of strong machines and weak wills. The film that captures this the best is *Wall-E*, which is a tale of two dystopias: broken Earth and boneless bodies. The unchecked pursuit of infinite desires turns the planet into a lifeless wasteland and it turns the people into listless blobs on a perpetual space cruise. It is an upside-down world where the robots are the lovers and the humans are cold. The robots are responsible and the people are immature. The Earth is alien and space is home.

Of course, the orthodoxy doesn't believe in such contradictions. It professes nothing but strength. It has a solution to the first problem (environmental collapse), which consists in exchanging dirty technologies for clean, green ones. The second problem is not a problem at all. The people may be slaves to their machines, but they are *happy* slaves. Life is *convenient*—safe, comfortable, and full of hyper-realities to titillate the neurons. In *Wall-E*, energy has done just what it is supposed to do: liberate, enrich, and disburden us. It is that vacation from humbler toil. This *just is* the "way of life" that we refuse to negotiate. It's a success, how could it be a problem?

It's paradoxical: success breeds failure. We cross a threshold. There is a phase change. For all our counting, we have no way to account for these threshold phenomena when more shifts from being better to being worse. Remember, we are in the age of hockey sticks. Our logic is strictly linear.

By the end of *Wall-E*, we are celebrating (of all things!) the relinquishment of space-age technology for a return to Earth. The ship's captain bends down with legs barely able to sustain his own weight to put a plant into the soil. His face beams with the virtue of good and simple things. At last, he feels his *own* capacity to do work. He remembers something about reality. He is no longer a mere appendage to the machines that in serving him *too well* ended up mastering him. Back from the stars, he has learned something about the human condition—that "For mortals, the 'easy life of the gods' would be a lifeless life" (Arendt 1958, p. 120).

The more we try to purify a simple picture of progress, the murkier it becomes. Good and bad start to arc and jolt in violent superpositions. This is what the ethics of our high-energy world looks like. Actions are both right and wrong. Shows of strength (like military parades) betray utter weakness (via dependence on foreign oil and global supply chains). Like the upside-down, we cannot ignore these full dimensions of energy any longer. Thinking them through is like sticking your finger in the socket. Beware: High Voltage!

The paradox is that the way of life we refuse to negotiate is itself pure negotiation. It is the creative destruction of accelerating change, which forces us to constantly adapt to a new way of being. This is the ethos of our high-energy culture. Faster, bigger, more. Go! Go! Grow! In just the last fraction of human history, we stumbled on fossil fuels like a bear hitting the jackpot with a hive brimming full of honey. We are marauding on a sugar high. "Our virtues," as Nietzsche said, are acceleration and infinity. We have adopted them so recently, yet we are convinced this has always been the best way to be human. It is so utterly weird but how quickly it becomes so normal!

Laura Ingalls Wilder's *Little House in the Big Woods* is set in the 1870s and the animal-powered world that Nietzsche knew. In the book, we meet "the horsepower." It is a machine for threshing wheat powered by a team of eight horses. The man who owns it travels between small farms and works with the farmers. His pay is usually a share of their crop. He stays at their house for a couple of days and is a guest at the table for supper. On Sunday, they rest and do little else than read the Bible. Laura's family calls the horsepower "The Wonderful Machine," because it saves so much labor. Laura's father loves the machine, which becomes part of his way of life. It fits into the rhythm like a new sax in the band.

Yet that machine participates in a more nebulous Machine. This one has no governor. It just grows and grows beyond all proportion, rhythm, and measure. The 'horsepower' becomes the mega-tractors of today that do not fit into that old, small-farm way of life. Indeed, they build entirely new ways of life around them. It's not a new instrument attuned to the existing band; it's a whole new act. The settler colonialists like Laura's father are also dispossessed. Their land too turns liquid and flushes them away. There is no settlement to be had! If you believed that your "way of life" would be the same as it was when you were a child, you are in for a rude awakening. Prepare to get angry at the very thing you refuse to negotiate. It is said that Nietzsche lost his sanity at the sight of a horse being whipped. We too have gone crazy with this incessant whipping!

This is the work of our proliferating upside-down world. It is the *measurelessness* that results when God himself has been electrocuted and no longer monitors the thin places. We are playing the role of god now. If you are in the orthodoxy, you will not see just how queer this is. You will call it 'progress' and you will not be able to imagine ever loving any other way of life. You won't understand the captain's urge to turn the ship around, back to Earth, to the real. To do so would be to commit the heresy of nostalgia at best, to be a reactionary at worst. When self-driving tractors take over, the sons of farmers will look back at their fathers with the same pity that man uses to gaze upon ape.

Philosophers have called this meta-machine or mega-machine of the upside-down by many names. Jacques Ellul (1964) called it *La Technique*. Heidegger (1977) called it Technology. He didn't mean this or that artifact. Rather, he meant technology as *poesis*, that which brings forth the truth. Technology reveals reality to us. The reality revealed changes over time. Modern technology is a particular kind of revealing that he called *das Gestell*. It is one "that puts to nature the unreasonable demand that it supply energy that can be extracted and stored as such."

Ben Franklin helped to invent the battery, but that's just a particular artifact. Heidegger's point is that the modern energy orthodoxy is the invention of Earth-as-battery. Like the man armed with a hammer who only sees nails, it sees standing reserves of energy everywhere. The Earth is *Bestand* or a collection of resources. For capitalism to take off (indeed for it to even become imaginable), we first had to believe nature was latent capital—just waiting for us to make it do productive work. We first had to believe that the Earth was a battery, a storehouse of E.

The wild Rhine River, like the Niagara River before the dam, is an engine out of gear. When we dam it, we bring it into its essence. It becomes a battery, Heidegger writes, in "the interlocking processes pertaining to the orderly disposition of electrical energy." Energy understood in this way is not just "the capacity to do work" it is the *command* that everything be made to work for us. All must be brought to heel in systems of control, regularity, and standardization. Everything, that is, must be *machined*. Heidegger notes, though, that 'we' don't issue this command—it is issued through us the way lightning raced through Franklin's kite string. This is how the world has been granted to us; it is our *doxa*. It seems natural and inevitable, but it is *not* our fate.

The mega-machine is on the move. First, we got slaves and animals to be our energy. Then we got steam engines. Then we got electricity. Then we got computers. Now we are getting artificial intelligence, AI. This last energy transition is when the upside-down creates its funhouse mirror image of humans, because AI will be able to do higher-order things like learning and planning. True acceleration will kick in and we will need life-long training in how to rapidly adopt a new way of life and then immediately dispose of it and forget about it.

Are we ready for these new intelligent energy slaves to have the capacity to do the kind of work we thought was reserved just for humans? Or will we need to modify church dogma to proclaim that some kind of work is for *Homo sapiens* only? Leave the intellectual labor to us! The fear in the earlier stages of industrialization (our progressive path to the high-energy way of life) was exploitation. Now the fear is obsolescence (Harari 2018). Who needs the humans in the right-side-up world if all they do is lounge around?

From the point of view of the mega-machine, we *Homo sapiens* are merely Humanity 1.0 (see Fuller 2011). If energy is just the capacity to do work, then we have to admit that the robots can *do far more* with this capacity than we can. Who needs our bodily virtues when the machines are so much stronger? Who needs the intellectual virtues of mere meat-bags—let alone their 'gut' instincts—when algorithms are so much better at making decisions? In this way, the orthodoxy (our one-eyed ethos) mistakes us for inefficiencies. We created this world, but are we gods or glitches?

The high-energy life cannot help but look like energy itself: restlessly shuffling around, taking on different forms. And of course *accelerating*! Yet we will defiantly insist that all these negotiations are not up for negotiation! This will continue even as we negotiate ourselves out of existence.

*   *   *

There was a time when our culture—call it "the West" for now—didn't house this dark mega-machine. When it was coined by Aristotle, *energeia* was not just a metaphysical but also a characterological term. It was about the upward oomph it takes to become an excellent human being. This is the other kind of energy Thoreau was talking about. As a matter of personal ethics, this requires channeling energies carefully to avoid burning out in a hedonic life devoted to lower-order pleasures. As a matter of collective ethics or politics, this requires restraints on techno-economic growth. In both cases, *limit* was essential.

In Plato's *Phaedrus*, Socrates makes the same point. He describes the human soul as a chariot with a driver and two winged horses, one of noble breed and one of ignoble breed. The charioteer has a "great deal of trouble" handling the horses as they aspire to the heavens where true pasturage can be found: "the vicious steed goes heavily, weighing down the charioteer to the earth when his steed has not been thoroughly trained." This is horsepower understood as soul force—the "painfully difficult business" of leading a good life. Notice the metaphor is about the ethical dimensions of *up and down*. Noble and base. It is the stuff of virtue.

But we live in an age "after virtue" (MacIntyre 1984). Especially as a collective political project, we have given up on trying to train that ignoble steed. As Nietzsche said, we have dropped the reins. No longer are "our virtues," as with Aristotle, a matter of "correct desire." We have stopped trying to reason together about the meaning of the good life or human excellence—about which desires are noble or right. The Amish may do this, but not us. We are moral strangers and have no frame of reference for deciding what *fits* and what does not. The operative energy now is not virtues but 'volts,' a universal currency that will bring you happiness no matter how you define it. To each his own! More power to you! Emotivism is the orthodoxy's moral theory—the good life is a matter of subjective preference. Whatever it is, more volts will get you more of it. As a result, we focus only on one kind of energy—the kind the orthodoxy knows so much about.

Despite its obsession with control, the central dogma of the energy orthodoxy is that people will *not* control their desires. Indeed, the whole system depends on people giving in to their desires. It will not tolerate governments that impose limits on consumption. Ergo, we must grow. *That* we simply cannot control. *That* is the one thing out of our hands. Behind all those numbers, those suits, those rigid machines, there is a throbbing hedonism. Oh, all the things we simply must have! We 'semi-barbarians'—it is as if we live in Maurice Sendak's book *Where the Wild Things Are*. The walls all around our boardrooms and bedrooms dissolve to put us in the company of our kindred, cavorting monsters.

The environmental scientist Jesse Ausubel captures this point succinctly:

> there is essentially little choice on a crowding planet to pursue technological solutions to feeding ourselves, shifting away from carbon-containing fuels, and otherwise limiting our ecological imprint. Human nature is probably harder to change than technology (in Revkin 2008).

Even the environmentalists are not going to give up their family vacations. They don't take Thoreau's call to 'simplify!' seriously. They read Thoreau, but they don't *live* him (as he wrote, "There are nowadays professors of philosophy, but not philosophers"). Ausubel has spelled out the orthodoxy's logic: just take desires as a given and find cleaner fuels to satisfy them. Like many, Ausubel thinks nuclear power is key to this strategy.

"The snake brain in each of us," he writes in his defense of nuclear power, "makes me cautious about relying heavily on changes in behavior. In contrast, centuries of extraordinary technical progress give me great confidence" that we can achieve "a prosperous, green environment." That "snake brain" is Plato's ignoble steed. For all its power, the orthodoxy has its origins in an act of surrender: *That* horse, the troublesome one, is just too strong to contain. We won't ever master the energies involved in making better people, in getting them to rise above their snake brains. But we can master the energies out there in the natural world. It is better to do battle with the forces of the atom than the horses of the soul.

This is our gamble. I wonder if it is wise.

Bibliography

Arendt, Hannah. 1958. *The Human Condition.* Chicago: University of Chicago Press.

Ellul, Jacques. 1964. *The Technological Society.* New York: Vintage Books.

Fuller, Steve. 2011. *Humanity 2.0: What It Means to be Human Past Present, and Future.* London: Palgrave.

Harari, Yuval. 2018. *21 Lessons for the 21st Century.* New York: Spiegel & Grau.

Heidegger, Martin. 1977. *The Question Concerning Technology and Other Essays.* London: Garland Publishing, Inc.

Heinberg, Richard. 2018. The New IPCC Report Offers Climate Solutions that Depend on Magic. *Pacific Standard*, October 8. https://psmag.com/environment/2018-ipcc-report-includes-magical-thinking

MacIntyre, Alasdair. 1984. *After Virtue.* Notre Dame: University of Notre Dame Press.

Magra, Iliana, et al. 2019. A Heat Wave Bakes Europe, Where Air-Conditioning Is Scarce. *New York Times*, July 25. https://www.nytimes.com/2019/07/25/world/europe/heatwave-record-temperatures.html

Revkin, Andrew. 2008. Can People Have Meat and a Planet, Too? *New York Times*, April 11. https://dotearth.blogs.nytimes.com/2008/04/11/can-people-have-meat-and-a-planet-too/

Smil, Vaclav. 2019. *Growth: From Microorganisms to Megacities.* Cambridge, MA: MIT Press.

Thoreau, Henry David. 1849 [1903]. *On the Duty of Civil Disobedience.* London: The Simple Life Press.

Wilder, Laura Ingalls. 1932. *Little House in the Big Woods.* New York: Harper.

4

From Virtues to Volts

That's not how the Force works!
Han Solo

The orthodoxy embraces the raw power of modern energy, but despises its superpositions. It resolves all tensions to one reality or the other. Schrödinger's cat, it insists, is either living *or* dead. Half-truths provide full assurance. The orthodoxy can confidently talk about 'energy' only because it has limited itself to half the picture. As an act of Socratic monkeywrenching, my focus on paradoxes is meant to keep the cat alive *and* dead. Though it makes things murkier by keeping us in a state of indeterminacy, we must attend to the energy of volts *and* virtues. If we don't, we will miss the energies that are harder to count but count the most.

There was a time in the nineteenth century, as the modern concept of energy was still being hammered out, that people used the term 'virtue' in ways that made this superposition clear. In his 1840 treatise on natural philosophy, for example, John Herschel wrote, "*there is virtue* in a bushel of coals properly consumed, to raise seventy millions of pounds weight a

© The Author(s) 2021
A. Briggle, *Thinking Through Climate Change*, Palgrave Studies in the Future of Humanity and its Successors, https://doi.org/10.1007/978-3-030-53587-2_4

foot high." After all, virtue can mean valor, the capacity to act, a potency, or power.

The kinds of energies signified by 'volts' have complex relationships with the kinds of energies signified by 'virtues.' But to oversimplify things for a moment: virtues are the energies of mastering the self and volts are the energies of mastering the world. The former is measured and proportioned, the latter is infinite and counted by metrics.

To get a sense of what I mean, picture master Yoda from *Star Wars* meditating in his hovel. He gets a knock on the door. It is the power company coming to do an energy audit. They look around briefly to discover in embarrassment that his hut is not even connected to the grid. The audit turns up dismally low numbers. From their point of view, Yoda suffers from extreme energy poverty.

Of course, as Han Solo noted, that's not what "the Force" is all about. Yoda personifies the kind of voluntary poverty, as Thoreau says, "that enjoys true wealth." There is actually an abundance of energy pulsing through Yoda's being—but we can only detect that if we have the right measuring tools. In the world of *Star Wars*, the Jedi are able to harness "the Force," which Obi-Wan Kenobi describes to young Luke Skywalker as "an energy field created by all living things." As we scarf down popcorn and slurp up soda in the movie theater, we intuit that there is something true about this other kind of energy and the higher nourishment it offers. We might also get the sense that had Yoda connected to the grid, bought a nice couch, and installed internet he would lose touch with the Force. Tuning in to higher energies takes habitual self-discipline.

This other kind of energy is complex and different cultures give it many names. It is what I will call simply 'virtue.' It is still not likely to strike the reader as a central term for a book about energy. That's because we are steeped in the orthodox ethos, which has tried to write virtues out of the cannon. Yet like those "thin places," virtues keep popping up.

Virtues and volts participate in relationships that can get out of whack. Even before Thoreau, the Scottish philosopher and social critic Thomas Carlyle diagnosed this problem of volts growing so big as to sap the energies of virtue:

Men are grown mechanical in head and in heart, as well as in hand. They have lost faith in individual endeavour, and in natural force, of any kind. Not for internal perfection, but for external combinations and arrangements, for institutions, constitutions, for Mechanism of one sort or other, do they hope and struggle. (Carlyle 1829)

An internal struggle becomes *externalized*. We go from grappling with our soul horses to grappling with machines and the social structures necessary to keep them running. Or rather, we've always done both, but something tipped that balance off kilter. As Ivan Illich (1974) puts it, the meaning of energy "was transmogrified from human vigor to nature's capital." Zhuangzi, successor to Laozi, tells a story about an old man who "used up a great deal of energy" to lug water from a well to his field. Someone asked him why he didn't go to the market to buy a machine that would ease his burdens. The old man replied, "where there are machines, there are bound to be machine worries; where there are machine worries, there are bound to be machine hearts" (in Watson 2013).

Of course, this energy transition to Mechanism is incomplete and messy. But it is real and worth at least as much of our attention as any transition from one mechanism to another. Indeed, it is more subtle than that. The struggle is always simultaneously internal and external. The trick is in the balancing.

Consider the example of the American aviator Charles Lindbergh. In 1927, at the age of 25, he made history with his nonstop flight from New York to Paris, covering 3600 miles in over 33 hours. Lindbergh helped to design the single-engine Ryan monoplane named *Spirit of St. Louis*. When he landed, he became the most famous man in the world. In the 4 months following his amazing feat, he visited 82 cities to deliver 147 speeches seen by 30 million Americans, a quarter of the total population. He earned the Medal of Honor, the nation's highest military decoration.

Medals are given for displays of virtue, in this case Lindbergh's ingenuity and especially his courage. The relatively small size of the engine allowed space for these excellences. As airplane engines grew over the following years, this space got squeezed out. The 'Mechanism' takes over to routinize the situation. By 2020, there were over 2500 transatlantic flights every day. The flight time from New York to Paris is seven hours. The experience becomes ever-more convenient, safe, and affordable. These are good things, to be sure, but they come at the price of excellence. No one gets a medal for what has become the routine service provided by those "external combinations and arrangements…[those] institutions." What was once virtuous would now be reckless, foolish, and unnecessary. And that is how the crank turns—the Machine closes in around those energies in us that yearn for heights.

Most histories of energy focus on the great and daring minds at the leading edge of innovation. But in her account of the oil boom on the Bakken Shale in North Dakota, Maya Rao (2018) gives us a people's history of energy. It is a story about the ruffians, ex-cons, and grifters of the oilfield. Her story makes plain one rather dispiriting fact about capitalism and modern technology. When the ideas are being birthed and the machines are only prototypes, there is the occasion for great feats of virtue and creative enterprising. But once the ideas are mature and the machines are mass-produced, the room for human excellence gets squeezed out. The genius casts his shadow over the masses. The inventor gives way to the man on the assembly line, and hordes of consumers ingest these bygone acts of daring with ignorance and impatience.

Once the elites of capital have uncorked the oily stomachs of the Earth all that is left is to labor and consume. The characters that fill Rao's books are the 99%. There is much to admire in them, but they are also made small by their setting, consigned to the various bit parts needed to keep the Machine humming. As the Machine grows, the cogs shrink and they are ground ever more finely by bureaucratic and market imperatives. They are the dim echoes of greatness, even if it is a destructive greatness.

The balance shifts. The orthodoxy will sell this to us as a story of liberation, but not all burdens are odious. Indeed, some provide the occasion for excellence. Many others provide the conditions for meaningful work. The biggest factor conspiring against coal miners, for example, is automation (Saha and Liu 2017). Long before climate change regulations and renewable energy could take a bite at that industry it had shed well over half of its work force. Enormous mining machines and computerized logistics enabled productivity to increase despite the job losses. In 1980, it was 1.93 short tons per miner hour. By 2015, productivity jumped to

6.28 tons per miner hour. Productivity and efficiency are the gold standards for the orthodoxy. They are good things, yes, but only *to a point*. It is the *measurelessness* of modern energy that bedevils us.

This often manifests as a crisis of male energies. Where, the men ask, is the moxie and brawn in the office job? And when that too is automated, where is the dignity in continually subjugating yourself to the Mechanism that always wins every arm-wrestling contest? It is emasculating. The same questions dog the progress of military weapons. Where is the virtue and valor of flying a drone in Afghanistan safe from a trailer in Las Vegas? And what does this do to the territory of the battlefield and the meaning of 'soldier'? Enemy combatants fall through thin places confessing an innocence that no longer exists. Ward Churchill (2003) had a point that office workers in Manhattan are complicit in the suffering of people far away. There are so many thin places. We are all complicit.

Like the labor of women that has so long been relegated to the shadows, men too are starting to find their energies to be unappreciated. Might this partially explain the spasms of nationalism roiling global politics? Yes, down with the patriarchy. But do we need to eliminate all places where men can be manly?

Do we dare remind ourselves of just how tough—how manly?—the pioneer women were? Have we grown too *soft* to make the *hard* decisions on our horizon when it comes to climate change? Instead, we might enact one of the core rituals of the orthodoxy: find a scapegoat, some beast to burden with the sins of others. Blame the migrants and climate refugees who have been displaced by the energies we disavow and who fall through the thin places we created yet refuse to acknowledge. This is the easy way out, and if our machines have taught us one thing it is always to seek the path of least resistance! It is only natural.

There are shifts from virtues to volts as machines grow bigger. Yet it is more complex than that. It is what Nietzsche called a revaluation of values. New ways of life with new virtues bloom and fade. Established virtues take on new meaning in the same old word. Where I live, there are many folks who drive big pickup trucks with bumper stickers professing their *conservative* values. A term that once signaled small scale, cautious self-reliance now signals utter dependence on a machine powered by vast technological networks spanning the globe. Nietzsche might call this a transition from a master morality to a slave morality. Yet the truck hides this revaluation, because we get the illusion of mastery with so many horses under our right foot.

*   *   *

Aristotle foresaw the way things can get out of balance. Indeed, his ethical and political writings are efforts to keep our current state of *measurelessness* from happening. One is not free, Aristotle argued, if one is a slave to desire or shackled by new needs. This was long an orthodox view. For example, four hundred years later, the Roman slave-philosopher Epictetus wrote: "For freedom is acquired not by the full possession of the things which are desired, but by removing the desire" (Higginson 1865, p. 210).

For the ancients, the purpose of liberation from necessity was not to acquire conveniences and amenities. Aristotle writes in his *Politics*, "household management attends more to men than to the acquisition of inanimate things, and to human excellence more than to the excellence of property which we call wealth" (1259b18, for all references to Aristotle see Barnes 1984). Yes, Aristotle thought there were "natural slaves." The purpose of slavery, though, was to grant the freedom and time needed to engage in contemplation as well as speech and action in the public sphere. Slavery wasn't intended to make it easier for the masters to gratify their desires. It was intended to give the masters a chance to truly become excellent.

This is *not* to justify slavery. And it is not to deny the ugliness of this facet of Aristotle's thought or to deny that slavery in ancient societies also often served decadence and trivial amusements. But it is to call out the key distinction that Arendt notes between ancient and modern forms of slavery. In antiquity, the institution of slavery "was not a device for cheap labor or an instrument of exploitation for profit but rather an attempt to exclude labor from the conditions of man's life" (Arendt 1958, p. 84). Labor and consumption are the stuff of animals, not humans. The master is to be free altogether *from* acquisitiveness not free *to* acquire more and more. Modern colonial slaves and machine slaves, by contrast, serve a freedom understood by the terms of productive labor, namely consumption and the generation of material wealth.

Here is the revaluation of values. Labor for Aristotle was our most beast-like quality. It kept us from truly human forms of excellence. For the modern energy orthodoxy, by contrast, labor is our most human essence. Arendt notes that we have become a society of laborers, which is to say consumers. Aristotle always insisted that to be truly human meant possessing appetites thoroughly infused with thinking; the thinking contributes right reason and the appetites contribute right desire. Mere consumption based on the pursuit of any-and-every desire is a debasement of the human condition, one that mistakes enslavement-by-desire for freedom.

The anthropologist David Graeber also struck upon this point in his exploration of debt in the contemporary United States. Noting how many people are effectively renting themselves to employers to do jobs they don't like in order to pay off their debts related to forced over-consumption, he concludes, "We've managed to take a situation which most people in the ancient world would have recognized as a form of slavery and turned it into the definition of freedom" (Graeber 2011). Freedom *from* labor and consumption has become freedom *to* labor and consume. But this latter kind of 'freedom' is chained to ever-growing needs and desires.

For Aristotle, there was no such thing as a distinct realm called the economy. To use Karl Polanyi's terms, the economy was 'embedded' in complex familial, social, religious, and other relationships. What I have been calling a revaluation of values (the shift from virtues to volts), Polanyi (1944) calls "the great transformation." It was not just the development of new sources of power or machines, but the dis-embedding of economic production from all other modes of interaction. This dis-embedding is the same rupture that Arendt calls the unnatural growth of the natural. Once dis-embedded, the market becomes an independent process governed by its own laws.

The scarcity that mattered for Aristotle was *not* due to the miserliness of nature, which provided easily enough for our sustenance. Further, our physical needs are limited and objectively determined. To accumulate material wealth past these limits is to misconstrue the good life. That life is spent in contemplation or in service to the polis where a different kind of scarcity reigns: the highest honors and rarest distinctions. These are scarce for the logical reason that if everyone were to have them, they would lose their meaning and function (trophies for participation ring hallow). We are on the planet to achieve *arête*, excellence, not to inhale as many goodies and experiences as possible.

Aristotle treated economic subjects as embedded in considerations of the whole: everything was related to community, justice, and self-sufficiency. The *oikos* or household was comprised of relationships, not productive activities or commodities. In the nascent profit-making buying and selling arrangements of the *agora* (the marketplace), Aristotle saw an existential threat to community. If markets started setting their own prices and the terms of trade, then unlimited growth and instability would follow. Aristotle, thus, designed a moral-political theory to demonstrate that human wants and needs are proportioned rather than boundless.

Central to that theory are the notions of "natural trade," which restores self-sufficiency to a community and "just prices," which strengthen the bonds of community. The key is that both concepts hinge on the notion of *limit* and strictly subordinate economics to communal welfare. Aristotle writes, for example, that "retail trade is not a natural part of the art of getting wealth; had it been so, men would have ceased to exchange when they had enough" (*Politics* 1257a17–18). Note the objective way in which 'enough' is used. Barter or *metadosis* is about mutual aid and sharing until all members of the community are brought up to the level of sufficiency. Aristotle says that "proportionate reciprocity" holds the community together.

Later thinkers like Leopold Kohr and E.F. Schumacher would develop this idea of proportionality. It can be put in terms of bodies. Living organisms possess a certain proportionality that can be expressed, for example, in terms of their surface area to volume ratio. A mouse can only exist within a certain range of sizes. If it were to keep growing bigger and bigger, it would need bigger limbs that *at some point* start to be rather un-mousy or un-mouse-like. Proportionality is what bodily health is all about. For example, consider the healthy range for a body mass index, a healthy range of sodium intake, and so on. Aristotle and these later theorists extrapolate from this physiological insight to social morphologies, or accounts of how human communal life must also observe certain proportionalities lest it be thrown out of whack.

Aristotle noted that the Graces have a prominent place in the temple as a reminder to do one's part. This is about kinship, friendship, honor, obligation, and the keeping of a right balance. There may be numbers in the scrolls that record bartering activity, but these are governed by a moral limit and are not the same as the prices established by the impersonal workings of an autonomous and invisible hand. This embedded world of mutual aid and sufficiency might give us a clue about the kind of world a new generation of climate activists is building. More on that in a later chapter.

*   *   *

For now, let's get back to that mother of all energy transitions—not from coal to wind, but from virtues to volts. Desires and needs are the prime movers of the human condition. They are what energy looks like from the inside of life. And they are tricksters. It can be hard to know what we really want or what would really bring us fulfillment. This is why philosophy has long ruminated on just what energies are driving us.

Consider two theories of desire, one ancient and one modern. Call them *eros* and electricity. Plato thought there was an upward direction and a proper target for *eros*. For Sigmund Freud, by contrast, it is about hydraulics. The libido is an energy that builds up pressure in the plumbing system of the body in the same way flows of electrons follow the pressure gradients of voltages. For Freud, desires, like electrons, have no proper aim. They are just seeking release. *Eros* and electricity, then, is another way to cast the difference between virtues and volts.

Those who wish to lay the blame for climate change at the feet of capitalism might be tempted to turn to Karl Marx for answers. Certainly Marxism paints a very different picture of society. Yet it is still on the 'volts' side of the energy transition: it is still an ideology premised on humans as laborers and consumers. It does not question the logic of growth built into the energy orthodoxy—it just wants to distribute the fruits of growing productivity more evenly.

Marx saw in industrial capitalism both the vastly increased productivity of "labor power" and its exploitation by the capitalists. Once his communist revolution took hold, he reasoned that *everyone* would be liberated from drudgery. What would they do with this freedom (this "vacation from humbler toil")? Well, they would engage in 'higher' activities, basically recreating an ideal Greek democracy without the elitism—one where all people engaged in civic virtues. Arendt, writing about a hundred years later, summarized what had become of this utopian vision. "We all know," she observed with dry wit, "the fallacy of this reasoning." The spare time created by modern energy services "is never spent in anything but consumption" (1958, p. 133).

Marx didn't recreate a more democratic golden age of Athens, because once our human energy is understood as labor power rather than *arête* or *eros*, we are bound to channel that energy into laboring and consuming rather than excellence of speech, action, or contemplation. As Arendt put it, because a laboring society levels everything, preferring quantitative equivalencies to qualitative ranks, "there is no class left, no aristocracy of either a political or spiritual nature from which a restoration of the other capacities of man could start anew" (p. 5).

In his history of liberal democracy, Patrick Deneen (2018) makes this point in a different way, lamenting "the submission of all forms of cultural life to the sovereignty of technique and technology." Think, for example, of the homogenizing sprawl of suburbia or the standardization of tests in education or the domestication of adventure into busses full of tourists. There is something pressing everything into its mold.

'Volts' is orthodox energy. It invites us into a world of professional discourse. It is the stuff of business, policy, and engineering. Volts are quantitative phenomena that can be converted through the International System of Units (SI) of mass, length, and time into the other metrics of serious energy discourse such as joules (J), watts (W), BTUs (1 BTU = 1055 J), and calories (1 cal = 4.18 J) (see Smil 2006). So, to say 'volts' is really to speak of energy as understood by the modern scientific worldview. The orthodoxy has physics for its foundation, which is the biggest reason it dominates our understanding of energy and its relationship to the good life.

On the other hand, 'virtues' conjures a discourse that is far more suspect. Is this kind of talk not outdated? And subjective? You can't measure virtues or calculate and convert them with precision, and any discussion about them will bog down in disagreement. Aristotle talks about having 'enough' as if there was an agreed-upon definition for that! Again, we have abandoned virtue as a sensible political rubric. Volts can bind us together, because they only ask us to agree to disagree. You use your volts one way, I use mine another. Give me prosperity, not propriety.

Our bias, in other words, is to accept some ways of speaking about energy as legitimate and to dismiss others. William James used his 1906 presidential address to the American Philosophical Association (APA) to challenge this orthodoxy when it was still congealing (see Bordogna 2008). His address was titled "The Energies of Men." James noted how the new scientific psychology had become prominent and respectable, but couldn't give an account of energy as people experience it in their lives. For example, we have all experienced a "second wind" where in a state of exhaustion somehow a burst of energy, rather than fatigue, takes hold of our being. And we all seek to tap into the higher energies of spiritual transcendence. James noted how scientific psychology ignored these connotations of energy, thereby relegating them to the quacks, mystics, crystal healers, and other pseudo-scientists.

Yet this is to ignore the most vital energies of all. What is the point of a human life if not to let our fires burn as hot and bright as possible? James was critiquing the philosophical naturalism of the orthodoxy, the view that science gives the only true account of reality. The world of SI units, though undoubtedly real, is derivative from and only made meaningful in the context of our lived experiences. James said that we need a science that takes the whole gambit of energies seriously:

> Everyone knows on any given day that there are energies slumbering in him which the incitements of that day do not call forth, but which he might display if these were greater... Compared with what we ought to be, we are only half awake. Our fires are damped, our drafts are checked. We

are making use of only a small part of our possible mental and physical resources. (Bordogna 2008)

Those 'fires' are part of what I want to capture with my use of 'virtues.' How do we lift the damper from our fires? That's the question that should guide our energy R&D programs.

The orthodoxy has an answer to this question: more volts = more virtues. The philosopher Carl Mitcham and the anthropologist Jessica Smith Rolston (2013) call the orthodoxy a "type I energy ethics" or a "pro-energy ethics." It preaches that access to more modern energy services (volts) will unlock more of our human potentials or capabilities (virtues). For the orthodoxy, volts are the enablers for virtues understood as capacities. They unlock our potential. Again, though this is true, it is a half-truth.

Thoreau flipped the orthodox formula on its head to arrive at the equal and opposite half-truth: more volts = fewer virtues. Like James (and Yoda), he was concerned with the energetic question of how one truly becomes 'awake.' He was looking for the key to those rare luminaries who are capable of the true exertion of the "poetic or divine life." The key: "Simplify, simplify." Less material development, more spiritual development. In *Walden*, he writes, "We are conscious of an animal in us, which awakens in proportion as our higher nature slumbers." There is our worm energy and our divine energy and they are inversely related. We get lazy and seek "to make this low state comfortable and that higher state to be forgotten."

The Breakthrough Institute report, *Our High-Energy Planet*, offers a good illustration of the orthodoxy. It opens with an assertion of self-evidence: "The relationship between access to modern energy services and quality of life is well established" (Caine et al. 2014, p. 4). They mean that better lives always follow from more modern energy services.

But it is not that evident or straightforward. Neither is it as simple as reversing the formula with Thoreau to conclude that better lives always follow from *fewer* energy services. Rather, we need to consider the perils of both too little and too much. This is a matter of paying attention to the way different kinds of energies interact with one another and seeking their proper balance. We have to be attuned to a phenomenon the orthodoxy cannot register: thresholds. A threshold is the toggling moment of the superposition, that arcing from pole to pole. It's the golden mean.

Smil, as we'll explore later, has shown that there is a threshold beyond which increasing energy consumption fails to deliver increased well-being. He was following in the footsteps of the heterodox Catholic thinker Ivan Illich (1973), who talked about superpositions as 'counter-productivity.' Energy slaves work for us, but only up to a certain speed or size, beyond which they become our masters and we start working for them.

To account for virtues means thinking about progress in terms of non-linear functions. We need to consider horizontal asymptotes (diminishing rates of return) and parabolas where more is better only *up to a point* (the sweet spot) beyond which more is worse. We need to think with new images and shapes. This is starting to take hold even in economics with a new generation of theorists challenging the old certainties of linear growth. For example, Kate Raworth (2018) pictures the new truths of the global economy in the shape of a doughnut. The ring of the actual doughnut is the sweet spot above the hole in the middle of the doughnut (representing deprivation) and the outer edge of the doughnut (representing excess) (Fig. 4.1).

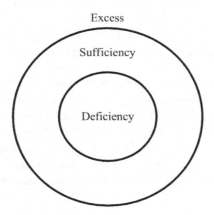

Fig. 4.1 The shape of virtue

In Buddhist terms, we need to aim for the "middle way." Confucius put this in terms of the "doctrine of the mean." Similarly, Aristotle understood virtues as the means between extremes.

This is how progress starts to look once we expand our sense of 'energy' to mean more than just the stuff that is delivered to us across wires and through pipelines. Orthodox proponents of development regularly say that poor people lead lives without energy and that we need to give them energy. Without dismissing the benefits of development, I suggest this is a dangerous way of thinking. It alienates us from the energies discussed by James, Yoda, and Thoreau. That is to say that rich people too suffer from a kind of energy poverty.

Bibliography

Arendt, Hannah. 1958. *The Human Condition*. Chicago: University of Chicago Press.

Barnes, Jonathan, ed. 1984. *The Complete Works of Aristotle*. 2 vols. Princeton: Princeton University Press.

Bordogna, Francesca. 2008. *William James at the Boundaries: Philosophy, Science, and the Geography of Knowledge*. Chicago: University of Chicago Press.

Caine, Mark, et al. 2014. *Our High-Energy Planet: A Climate Pragmatism Project*. The Breakthrough Institute, April. https://s3.us-east-2.amazonaws.com/uploads.thebreakthrough.org/legacy/images/pdfs/Our-High-Energy-Planet.pdf

Carlyle, Thomas. 1829. Signs of the Times. *Edinburgh Review*, no. 98. https://salemcc.instructure.com/courses/451/pages/thomas-carlyle-signs-of-the-times-1829

Churchill, Ward. 2003. *On the Justice of Roosting Chickens: Reflections on the Consequences of US Imperial Arrogance and Criminality*. Chico: AK Press.

Deneen, Patrick. 2018. *Why Liberalism Failed*. New Haven: Yale University Press.

Graeber, David. 2011. *Debt: The First 5,000 Years*. New York: Melville House.

Herschel, John Frederick William. 1840. *A Preliminary Discourse on the Study of Natural Philosophy*. London: Longman, Rees, Orme, Brown, & Green. http://darwin-online.org.uk/content/frameset?itemID=A276&viewtype=text&pageseq=1

Higginson, Thomas Wentworth, trans. 1865. *The Works of Epictetus*. Boston: Little, Brown, and Company.

Illich, Ivan. 1973 [2001]. *Tools for Conviviality*. London: Marion Boyers.

———. 1974. *Energy and Equity*. New York: Harper & Row.

Mitcham, Carl, and Jessica Rolston. 2013. Energy Constraints. *Science and Engineering Ethics* 19: 313–319.

Polanyi, Karl. 1944. *The Great Transformation: The Political and Economic Origins of our Time*. Boston: Beacon Press.

Rao, Maya. 2018. *Great American Outpost: Dreamers, Mavericks, and the Making of an Oil Frontier*. New York: Hachette Books.

Raworth, Kate. 2018. *Doughnut Economics: 7 Ways to Think like a 21st Century Economist*. White River Juncture: Chelsea Green Publishing.

Saha, Devashree, and Sifan Liu. 2017. Increased Automation Guarantees a Bleak Outlook for Trump's Promises to Coal Miners. *Brookings*, January 25. https://www.brookings.edu/blog/the-avenue/2017/01/25/automation-guarantees-a-bleak-outlook-for-trumps-promises-to-coal-miners/

Smil, Vaclav. 2006. *Energy: A Beginner's Guide*. Oxford: Oneworld Publications.

Watson, Burton, trans. 2013. *The Complete Works of Zhuangzi*. New York: Columbia University Press.

5

Trespassing

It is only when you take your ethics for granted that all problems emerge as problems of technique
Louis Hartz *1955*

I was arrested on June 1, 2015. The charge leveled against me and a pair of my friends was trespassing. You might say this makes sense because we were on private property without permission. But there is a catch that I will get to momentarily.

The property in question was the entrance to a fracking site in our hometown of Denton, Texas. We sat in the middle of the driveway so that the big trucks full of sand and chemicals couldn't get past us on their way down to drill yet another gas well. Rather than squish us, they were nice enough to call the police. The police tried to reason with us, but we were determined to sit there. We were polite in our refusal to leave and they were polite in their insistence that we go.

The birds sang in a row of trees in the neighborhood across the street. It was a lovely morning. We sat there a while and everyone else stood there a while. But time was wasting and there was work to be done. So,

© The Author(s) 2021
A. Briggle, *Thinking Through Climate Change*, Palgrave Studies in the Future of Humanity and its Successors, https://doi.org/10.1007/978-3-030-53587-2_5

they put us in handcuffs. And while we sat in jail, fracking commenced at the site known as "Long Term Baldinger," because the minerals are owned by Brian Baldinger, a former National Football League (NFL) player who lives in New Jersey. You can watch him on cable television with other people in nice suits who find a way to talk about football every minute of every day of the year.

Denton has over three hundred gas wells, but this one was special. This was the first site that was re-opened after the Texas legislature overturned our municipal ban on fracking. The legislation, known as HB 40, said that any local ordinance on gas well development must be "commercially reasonable." Our ban did not pass that new test. But it had passed another kind of test: it had been enacted through a democratic process and approved by popular vote in November 2014.

Here is where the catch comes in. Were we engaged in an act of civil disobedience protesting the HB 40 state law? In that case, the arrest would be appropriate. Or were we actually enforcing our own city fracking ban, a law that was still on our municipal code of ordinances at the time? In that case, rather than getting arrested we could have, say, been awarded a medal or something.

We talked about this particular paradox with the police before we were arrested, but it got harder to think it through when one of the truck drivers starting getting angry and blasting his horn at us. The conversation ended in a muddle. Before the handcuffs came out, the Sergeant shook our hands as a way of thanking us for trying to protect our city from a uniquely invasive industry. I took that as an acknowledgment that although we were on the wrong side of the law, the law was on the wrong side of right.

The Denton fracking ban had been the culmination of a years-long process. At first, it was collaborative as residents worked with city officials and industry representatives to craft rules to protect health, safety, and neighborhood livability. But as I chronicle in *A Field Philosopher's Guide to Fracking*, things became acrimonious and polarized. No consensus could be found, so it came down to a vote. We won that battle, but with HB 40 we lost the war.

Then a funny thing happened. Shortly after our arrest, the market effectively banned fracking in Denton. The glut of natural gas from the industry's success drove prices down and it was no longer profitable or "commercially reasonable" to keep fracking. The drilling rig count on the Barnett Shale dropped to zero. The whole hullabaloo quickly died down. No more protests or petitions.

Energy extraction was where everyone seemed to want it: out of sight and, thus, out of mind. Life went on just like it had before. People stared at Facebook on their phones. They drove around to buy laundry baskets at Walmart. University Avenue expanded from four lanes to six. The old Rayzor Ranch on the west side of town got eaten up by a marvelous development full of shops and restaurants. You could buy a new cell phone at one store and play with it next door at the Whataburger. And if that wasn't satisfying enough you could get an ice cream cone or catch a show at the new cinema.

Life, in short, was good.

One day, while eating my ice cream and watching the NFL network, it dawned on me that underneath even that bitter fracking fight there was a deep consensus. Nobody questioned the high-energy way of life. Those opposed to urban fracking were not opposed to *that*. We didn't question the ends—we only questioned the means. It was, in other words, a feud between two factions *within* the orthodoxy. Give us the ice cream and all the other conveniences of modernity, just don't put the ugly business of energy extraction in our backyard.

The orthodoxy is full of debates about justice, that is, about how to distribute goods. But it almost never features debates about what to count as 'good' in the first place. There are many ways to police this boundary between justice and the good life. In the fracking fight, we did so by telling ourselves that this was all a matter of *land use* policy, not *energy* policy. Energy is fine. No one has a problem with it. It's just that we don't put bakeries in neighborhoods, why fracking sites? We weren't the trespassers. The industry was the trespasser intruding on our quiet lives. It should go somewhere else!

But where? Those quiet lives, those unquestioned goods, are made possible by the ugly, noisy business we protested. Why shouldn't it be in our backyard? Why shouldn't we be constantly reminded of just what it takes to make our way of life possible? If you want the burger, then go to the slaughterhouse and look it in the eyes. Those who consume the most should build their mansions on trash heaps.

Yes in fact we *are* the trespassers. The richer we become the more our way of life spills over. We stamp our non-negotiable pattern of life everywhere. We send tentacles across the globe where they bathe in the blood-oil of the Middle East and crack the whips in Asian sweat shops and strip mine the Andes. It's all there in the walls of Whataburger, the asphalt under the tires, and the computer chip in the phone. Even the electric vehicle carries in it lithium from a mine that leaked toxins into a Tibetan river (Katwala 2018).

It's all there. And, of course, it is not there. Look at the walls again, the tires, the phone. There's no injustice to be seen. It's just a lazy Sunday afternoon. What crisis? This is the metaphysical conundrum I have been calling superposition. Like Schrödinger's cat that is simultaneously alive and dead, we are simultaneously guilty and innocent.

* * *

The ethics of energy is about trespass. Energy just *is* the conversion of this into that—electrical into mechanical, solar into chemical, and so on. It is pure boundary crossing. The more energy we intake, the more we participate in the interloping. Yet, paradoxically, the more isolated we become.

High energy simultaneously individuates *and* collectivizes us. Disembedded from localities and traditions, the modern individual emerges as the sovereign center of experience and agency. At the same time, though, the modern individual is tethered ever more tightly to the emerging human-planetary collective. This is all captured in the experience of sitting alone in your car in a traffic jam: simultaneously the captain of your own ship and totally at the mercy of larger forces. We find ourselves often feeling insulated and isolated, but at the same time we are implicated and complicit in events around the globe. Schrödinger's thought experiment led him to coin the term *Verschränkung*, entanglement.

The first command of ethics is "know thyself!" But how can we know a self that is both hither and thither? We are hither-thither beings, stitched and sewn here and there.

In 1790, Allesandro Volta zapped the femoral nerve of a dissected frog leg. The damn thing jumped around even though the frog proper was dead and gone! The young Mary Shelley was paying attention when this kind of experiment was reported in the Republic of Letters. The new energy (what I am calling 'volts' for short) born in Volta's lab—and on Ben Franklin's kite string—was going to animate monsters. Not just one, but a whole army of Frankensteins would first dissect the world and then stitch pieces and parts together in new ways and shout "It's alive!"

Things have since been amped up and rearranged. On the right-side-up, all this energy looks like convenience. The stitching of monsters is the *convening* of pieces and parts. It is how the fruits of the Earth convene in your fridge or the labor of thousands of specialists convenes in your cell phone. Florida oranges appear on your kitchen table in Seattle. This is the world of the sovereign individual. Seen from the upside-down, though, all this energy looks like a "great derangement" (Ghosh 2017). Things may convene, but they no longer *gather* together in ways that rhyme. Places are ruptured and sutured somewhere else. Walruses fall into our living rooms, because our living rooms fall on them. Your electric vehicle is stitched to a dead yak floating on a river in Tibet.

This superposition is often couched in terms of "collective action problems." The most famous example is Garrett Hardin's 1968 essay "Tragedy of the Commons." Hardin imagines a group of herders who graze their animals on the commons. He then toggles across the poles of the superposition. Seen from the individual level, it is good to put more animals on the pasture. It returns more wealth. Seen from the collective level, though, this gives rise to over-grazing and, eventually, ruin for all. He argues that this describes many of our ethical dilemmas on a high-energy planet. The gasoline in your tank gets you to the grocery store so you can feed your family. Good. But it also contributes to air pollution and climate change. Bad.

Good and bad. The same act exists in two opposite states at once, just like the cat that is alive and dead. It depends on how you look at things. Again, the ethics of energy and the physics of energy both live in indeterminacy.

Hardin's solution is to limit individual freedoms to consume through "mutual coercion mutually agreed upon." For this, he has been widely attacked. Some argue that free markets can solve problems better than such heavy-handed government regulations. This often entails privatizing the commons, because people care for their own private property more than they care for the commons. In other words, assert the individual pole of the superposition as much as possible.

A more interesting way to critique Hardin is to revive, not eliminate, the commons. In other words, emphasize our collective identity as commoners pursuing a common good. Hardin had a dark take on the common good. He advocated for "lifeboat ethics," arguing that the Earth is like a lifeboat that can only sustain so many people (Hardin 1975). If we keep trying to pull people out of poverty, eventually the whole lifeboat will sink. Though it is brutal on an individual level, the collective good is served by refusing aid and development assistance to the impoverished. Some must drown, he reasoned, so that others can float along.

But isn't this immoral if not downright evil? In practice, it would amount to eco-fascist racism, xenophobia, and even genocide. The immigrants and refugees of the climate crisis who are coming to the United States and the EU are offering a test for our moral character. Do we really believe that our countries are full? Is the lifeboat at capacity already, despite our wealth and our ingenuity? Didn't we cast them adrift in the first place through the deranging forces of colonialism? A study sponsored by the US government predicted that in the face of an abrupt climate change scenario, wealthy nations will harden their borders "to hold back unwanted starving immigrants" (Schwartz and Randall 2003, p. 18). The moral derangement here is laying the blame for environmental ruin at the feet of the poor. Hardened hearts precede hardened borders.

So, emphasizing the common good requires expanding our empathy or our sense of self to identify with our fellow terrestrials. We are in this together. This implies an ethics of solidarity and mutual aid reminiscent of Aristotle's focus on community. Having your first thought be 'Carbon' in the morning means that you are thinking about all the ways 'you' are not confined to the here-and-now. What you are trying to overcome with this moral exercise is the illusion of the self that comes from all the magic of the high-energy life. It sure seems like you are alone in your air-conditioned home. No, you are an accomplice, meaning you are folded with and together. You are in the commons.

* * *

The commons today is not what it was before the "great transformation" or the revaluation of values from virtues to volts. It is not what it was before it was ripped apart and sewn back together. The destruction of the commons is known as expropriation or dispossession. Arendt argues that the expropriation that began with the Reformation is one of three central forces giving birth to the modern world (along with the 'discovery' of the Americas and the invention of the telescope).

The result of this process was the creation of laborers. The commoners became a new class of wage slaves with only their labor power to sell on the newly dis-embedded market. Their once stable world was liquefied and poured into the fuel lines of industry. This labor power, this newly discovered form of energy, unleashed into society and organized by capitalists is essential to the ever-expanding economy.

In "Silence is a Commons," Ivan Illich (1983a) notes the social and environmental 'disvalue' that results when the commons is transformed into a "productive resource." As services and commodities grow, they exclude non-market forms of human relationships. Commoners are rendered landless and dependent on paid work, a fate that had long been considered worse than beggary (Protestant thinkers such as John Calvin had to convince the newly expropriated working class about the nobility of labor). Policing and legal systems of surveillance, exclusion, and control displace informal relations based on trust, tradition, ritual, and mutual aid. And despite enormous productivity, modern economies do not solve the problem of scarcity. Instead, the economy actually produces scarcity through force, need, or envy. Meanwhile, the lords of capital convince us that poverty is not a systemic requirement but the personal vice of laziness.

In short, we have lost our sense of the common good. A clear indicator of this is the fact that the twenty richest Americans gave an average of 0.8% of their wealth to charitable causes in 2018. That may not matter so much if there was a tax system in place to redistribute wealth along some moral sense of the commonweal. But that's not the case in the United States, where taxes have gotten more regressive as the economy grows. Bezos, the wealthiest of them all, finds a way to pay zero taxes (see

Saez and Zucman 2019). Of course other developed nations do much better in this regard.

Expropriation is the deprivation of a group of people from their place in the world and their subsequent exposure to the exigencies of life. The resulting laboring class literally lives from hand to mouth. Arendt notes that this had happened in the past, but the modern expropriation is distinguished by the way the wealth created was transformed into capital through labor. There was not just a redistribution of wealth, but a growing process of productivity and further wealth creation as well as wealth concentration at the top of the pecking order.

In other words, the destruction of the commons created a new form of poverty that was necessary for the generation of wealth, because without desperate wage slaves there was no energy to get the engines of capitalism off the ground (see Perelman 2000). Adam Smith, for example, was frustrated by the laziness of the average commoner and the way they tended to saunter around. Yet this was just the pace of the lifeworld before the creation of an expropriated laboring class that had to hustle to survive. Much of the 'progress' achieved by the orthodoxy and its provision of modern energy services is only a gain when examined against the baseline of modernized poverty. In other words, once people are evicted from the commons they come to need the services provided by the economy, especially jobs. That is, the energy orthodoxy fixes lots of problems, but they are often problems of its own creation (Smaje 2015). And its need for an underclass (the imperative of cheap labor) is obvious in the way inequalities and contingent labor continue to grow along with productivity and wealth.

Energy is often defined as the capacity to do work, but the energy of labor power is actually antithetical to work. For Arendt, work constitutes the building of a durable world that stabilizes human life. Human-as-worker, *homo faber*, reifies (makes solid or real) an objective world that relate us in our subjectivities. This world is wrested from nature either by felling the tree that becomes the beam in the house or by interrupting natural processes as when coal, iron, or marble are "torn out of the womb of the earth" (Arendt 1958, p. 139). The world then houses and orients us as well as protects us from nature. It is the place of gathering as when we come to the kiva, or sit around the table, or listen to music around the old courthouse on the square.

But labor and the *animal laborans* created by expropriation start to overshadow work and *homo faber*. The former produces consumables, the latter makes use objects. As the division of labor, mechanization, and automation increase the speed of the laboring activity, even use objects like buildings start to be treated more like items for consumption. They lose their status as the durable anchors of a world that binds the generations together and orients new human life. The old ranch I mentioned above was converted almost overnight into buildings offering dining and shopping experiences. One of the roads wending through this brand new 'development' is ironically named Heritage Trail. Then again, maybe it is not so ironic, given that constant change is the only heritage we're permitted.

Arendt writes that human labor power is "perhaps even the most powerful of all natural forces." Just look at how, once it is unleashed, it devours not just the natural Earth but also the "'unnatural' and purely worldly stability of the human artifice" (p. 126). The result is "a waste economy, in which things must be almost as quickly devoured and discarded as they have appeared in the world" (p. 134). Durability, the marker of work, is "the greatest impediment to the turnover process, whose constant gain in speed is the only constancy left wherever it has taken hold" (p. 253). For a laboring society, everything must be a consumable. That means planned obsolescence is the name of the game.

The eclipse of work gives rise to what Arendt calls "world alienation," where people are deprived of a sense of a stable home. Instead of places for gathering, they are offered spaces of convening. They are given the dazzling goodies on display in the air-conditioned units at the new shopping center or, even more conveniently, the endless scroll of consumables on Amazon and other websites. They are asked to be nimble foxes darting in and out of the contingent spaces afforded by an economic process of creative destruction.

The orthodoxy frames all of this in terms of happiness, opportunity, choice, and a rising tide of growth that lifts all boats. But there is a yang to that yin; there is a dark and destabilizing energy also at work. Will the superficial, transient amusements offered by our laboring society be enough to hold back the angst induced by world alienation?

Hillary Clinton lost the three states of Pennsylvania, Wisconsin, and Michigan in the 2016 US presidential race, in part because of the disaffection of white, non-college educated men who have been harmed by the globalizing impulses of a laboring society. Donald Trump promised a return to the past, the time of *homo faber*, when they had a decent job, a steady paycheck, and a stable world. Clinton proposed a series of worker re-education programs. Yet as it turns out, "a lot of coal miners are not interested in becoming computer programmers or dental hygienists" (Frodeman 2019). Arendt would recognize the nationalism gripping her home country of Germany and many other parts of the world. In rebelling against a ceaseless, faceless, accelerating process of change, people may try to hold onto the solidities and certainties offered by blood and soil.

*   *   *

As the French thinker Bruno Latour (2018) writes, "Against globalization and against the return to national and ethnic borders" (p. 100). But if we are turning away from both the totalizing growth of a global economy and the tribalism of yesteryear, what are we turning *toward*? Against those things, but *for* what?

This is where our moral imagination needs to stretch. I felt something that day at the protest at the frack site as well as at the trainings leading up to it. It was camaraderie and solidarity. It was Aristotle's notion of political friendship—the kind of relationship that exists outside of both the government and the market. It was human contact unmediated by institutions and bureaucracies. It was the commons in its classic sense. Social movements like the global climate protests, I think, are trying to revive this classic commons in new ways. Latour would say that they are trying to push us to a new pole that is neither local (no such thing exists anymore) nor global (this is unsustainable) but Terrestrial. It is hard to imagine what this means.

That's because it pushes us outside of the Overton Window. The orthodoxy will not tolerate such blasphemy. The common good and solidarity are heretical beliefs for a system that insists it is in the business of liberating *individuals*. The *anthropos* for the orthodoxy is the man on the couch surrounded by a home entertainment system, *not* "the people" sitting on the road with arms linked staring down frack trucks. We can discuss *movement* in the sense of force, mass, and acceleration. Talk of emotionally *moving each other*, however, is not permitted. The same goes for *being moved* into action. We can transition from coal to solar, but we cannot transition to shared territory.

Or maybe we can. There is certainly ample precedent. The antidote to Hardin's pessimism is the Noble Prize-winning economist Elinor Ostrom (1992). She repeatedly showed how the so-called tragedy of the commons is only tragic for cultures that have lost any sense of the *common good*. When we think about the commons not as a resource but as a shared responsibility, things shift. It is true that many cultures have caused environmental ruin. Yet it is also true that numerous and varied cultures have found ways to live sustainably in the commons and with embedded economies. They have not needed 'coercion' by a faceless bureaucracy and they have not let the invisible hand sort things out. Rather, they have developed trust, self-determination, and 'polycentric' decision nodes that empower those closest to the issues to have a voice.

Ostrom's work, along with that of anthropologists and many others, helps us to imagine the stunning diversity of cultures and the many ways of being human. To see the Anthropocene as the work of 'humans' is what feminists or critical race theorists would call an act of *privilege* (e.g., McIntosh 1990). It is to hide cultural institutions of power behind the mask of nature. As if to say, "It cannot be helped. This is just who *Homo sapiens* is!" Meanwhile, of course, not all humans follow this 'natural' law of behavior. In other words, the privileged get to ignore their disproportionate contributions to the problem of climate change and the disproportionate impacts felt by the under-privileged. They lack the pain that might bring the empathy to walk in the shoes of others. The concept of the 'Anthropocene' can reinforce this privilege by casting climate change as a human problem (*all* humans equally) and hiding all the power, violence, and exploitation going on between groups of humans.

As the Swedish scholar Andreas Malm notes, "If some humans introduced steam power against the explicit resistance of other humans, then it would be hard to maintain a notion of it as the expression of a species-wide project" (2016, p. 36). He continues, "steam arose as a form of power exercised by some people against others" (p. 36).

Who has ripped apart and then stitched this world back together? It is not the Zuni on the Little Colorado River or the *zoon politikon* (political animal) of ancient Greece. It is not everyone equally. Not all humans can recognize themselves in the a*nthropos* of the Anthropocene. The climate refugees being evicted by sea-level rise and drought are not the guilty parties. The poorest 45% of humanity creates 7% of total CO_2 emissions, while the richest 7% produces 50% of total CO_2 emissions (Malm 2016). I consume thirty-five times the energy as the average person in Bangladesh.

For these reasons, it may well be best to call this age the Capitalocene (see Malm 2016; Moore 2015). Expropriation shows us that the central driver isn't coal or other fossil fuels; it is the power of the capitalists to use coal on a dis-embedded market to exploit the laboring class and to unleash a growing, accelerating process around the world that demands cheap labor power. That is, until robots dispossess the laboring class. The Capitalocene arguably started in 1602 with the chartering of the Dutch East India Company, the world's first global corporation.

The turn toward a new commons is hopeful, but it would have us trespassing way outside the boundaries established by the Overton Window. I keep thinking of the Amish as an example, and most people don't want to live as they do. When Aristotle wrote about the commons, the notions of proportionality and sufficiency were crucial. So, the old commons was governed by a metaphysics of finitude. There was agreement on what ENOUGH looked like. And the virtues were tightly linked to one's role in serving the common good.

To point out the obvious: these are no longer our virtues. The 'good' and 'justice' for us are no longer about fittingness. The individual is sovereign. Limits are to be transcended. For example, to talk of healthcare as a human right is to set us down the path toward immortality, because this right has no limit. We have a right now to life-saving medicines and operations that didn't exist fifty years ago. In the future, we will have rights to 3-D printed organs and whatever else is next. We are not going to stop. Our virtues are too large for any tailoring. We are trying to don a cosmic, not an earthly, robe. This is why the orthodoxy is preparing to launch us off the planet, to another floating ball of resources.

Here are the cold, harsh words of the orthodoxy: "we are not in the commons in any meaningful way. We are, rather, on the market. The ethical imperative is not sharing but growing. We need to keep growing the pie so that everyone can get a slice."

The energy transition we have been tracing has brought this shift from sharing to growing. It is the revaluation of our values, our sense of justice and the good life. We are back to square one: can this be sustained? Because the slices of the pie keep getting bigger as the pie grows. No one, even the rich, ever seems to have *enough*. The tentacles may slither from coal mine to lithium mine as we transition from internal combustion to electric vehicles. But the tentacles remain greedy grabbers. And this is *good*. If Bangladeshis "catch up" to my level of consumption, they won't be blamed—they'll be applauded. Congrats: you made it to the good life! And we in the so-called developed world will pat ourselves on the back. Congrats: justice has been served!

Thus, the paradox stated differently. Because we understand the good life as consumption and we understand justice as consumption-for-all, we are morally obligated to do the immoral: to help everyone into a way of life that destroys the planetary basis of civilization. We have chained justice to endless growth. It is true "by every available canon of distributive justice" that the poor of the global south who have been deprived are entitled to their share of the economy (see Ghosh 2017). Even justice with her scales has tipped out of balance and participates in the monstrous unnatural growth of the natural. As the Indian writer Amitav Ghosh puts it, "Our lives and our choices are enframed in a pattern of history that seems to leave us nowhere to turn but toward our self-annihilation" (p. 111).

Bibliography

Arendt, Hannah. 1958. *The Human Condition*. Chicago: University of Chicago Press.

Briggle, Adam. 2015. *A Field Philosopher's Guide to Fracking: How one Texas Town Stood up to Big Oil and Gas*. New York: Liveright.

Frodeman, Robert. 2019. *Transhumanism, Nature, and the Ends of Science.* New York: Routledge.

Ghosh, Amitav. 2017. *The Great Derangement: Climate Change and the Unthinkable.* Chicago: University of Chicago Press.

Hardin, Garrett. 1968. The Tragedy of the Commons. *Science* 162 (3859): 1243–1248.

———. 1975. Lifeboat Ethics. *Hastings Center Report* 5 (1): 4.

Hartz, Louis. 1955. *The Liberal Tradition in America: An Interpretation of American Political Thought Since the Revolution.* New York: Harcourt Brace.

Illich, Ivan. 1983. Silence Is a Commons. *CoEvolution Quarterly*, Winter.

Latour, Bruno. 2018. *Down to Earth: Politics in the New Climatic Regime.* Medford: Polity Press.

Malm, Andreas. 2016. *Fossil Capital: The Rise of Steam Power and the Roots of Global Warming.* London: Verso.

McIntosh, Peggy. 1990. White Privilege: Unpacking the Invisible Knapsack. *Independent School*, Winter, pp. 31–36.

Moore, Jason. 2015. *Capitalism in the Web of Life: Ecology and the Accumulation of Capital.* London: Verso.

Ostrom, Elinor. 1992. Institutions and Common-Pool Resources. *Journal of Theoretical Politics* 4 (3): 243–245.

Saez, Emmanuel, and Gabriel Zucman. 2019. *The Triumph of Injustice: How the Rich Dodge Taxes and How to make them Pay.* New York: W.W. Norton.

Schwartz, Peter, and Doug Randall. 2003. An Abrupt Climate Change Scenario and Its Implications for United States National Security, October. https://eesc.columbia.edu/courses/v1003/readings/Pentagon.pdf

Smaje, Chris. 2015. Dark Thoughts on Ecomodernism. *The Dark Mountain Project*, August 12. https://dark-mountain.net/dark-thoughts-on-ecomodernism-2/

6

I Kant Believe You

Dare to know!
Immanuel Kant, "What is Enlightenment?" 1784

Energy enervates. Somehow, the very forces that have given us the muscle to re-shape the entire planet have also sapped our strength or, better said, our *will*.

The teenage climate activist Greta Thunberg has become the face of a new generation's fight against climate change. Thunberg gained notoriety for helping to start the global "school strike for climate" movement. She has addressed the UN, the EU, the US Congress, and other major political bodies. She's been nominated for the Noble Peace Prize and featured on *Time* magazine. Thunberg is on the autism spectrum, and she is right to suggest that the so-called normal adults are the ones with the disability, not her. She asks why carbon dioxide emissions keep going up and why species keep going extinct:

> Are we evil? No, of course not. People keep doing what they do because the vast majority doesn't have a clue about the actual consequences of our everyday life, and they don't know that rapid change is required. We all

© The Author(s) 2021
A. Briggle, *Thinking Through Climate Change*, Palgrave Studies in the Future of Humanity and its Successors, https://doi.org/10.1007/978-3-030-53587-2_6

think we know, and we all think everybody knows, but we don't. (Thunberg 2019)

We know it, but we don't really *believe* it. That is to say that we don't have the will—the truly scarce energy resource. We know but we don't *dare*. We understand the game being played but we don't put any skin in it.

The French engineer-philosopher Jean-Pierre Dupuy (2009) also notes how our psychic energies are at the heart of our energy-climate crisis. Psychologically, he argues, we only believe a crisis exists if there is a solution. To say that we must change our whole way of life strikes most people as wholly untenable so they do not see the crisis. Moreover, a catastrophe only becomes possible by possible-izing itself, by becoming real. But by then, it is too late (and the catastrophic becomes the new normal). To prevent the catastrophe is to keep it impossible, which means it won't seem real, which makes it harder to prevent.

Dupuy puts this in terms of another paradox. Usually in ethics, *ought* implies *can*: I have a moral obligation to do something only if I can do it. But in this case we are morally obligated to do something we cannot do, namely, to foresee the future. This is impossible, because the systems in play are too complex. Doing scientific research often just adds to more uncertainty and resulting paralysis (Sarewitz 2004). This situation, Dupuy argues, calls for a new metaphysics of time, where the future is made real in the present. The time-traveler who does this is the *prophet*, one who speaks with clarity about a future that is simultaneously real and unreal. Thunberg and the other young climate strikers are the prophets of our age, but with a twist. Prophets have historically had a performative power: by saying things they bring them into existence. These climate prophets are saying things about a devastated future, a future that already exists, in order to take it out of existence.

When Thunberg walked out of school to strike for climate action, some people told her she should stay in school to become a climate scientist to "solve the climate crisis." In her 2018 TEDx talk at Stockholm, she puts the phrase in condescending air quotes because scientific knowledge isn't the solution. We are drowning in facts and numbers. Everyone has their own bags full of those. "All we have to do," she says, "is wake up and change." It is indeed as if we are sleeping. Or worse: drunk at the wheel.

The ethical dilemma Thunberg has spotted is not only pressing, it is perennial: how can we *know* but not *act*? Socrates believed that knowledge and virtue are the same, because one who *knows* what is right will also *do* what is right. But that doesn't seem to be the case. What's going on? Socrates says that a man might know that getting drunk is not wise, but "when the impulse comes upon him, he may turn his attention away from this knowledge." When he surrendered to the impulse, the "knowledge had slipped from the field of his mental attention" (Copleston 1993, p. 110). And what does high energy do to us if not constantly play to our impulses!

Our "field of mental attention" is one giant sieve, always letting knowledge slip through. Haven't you been reading the dire news about biodiversity loss on your phone during the commercials? And then you turn it off to laugh along with the show. Look there at the check-out aisle in the grocery store: fifty brands of candy for our impulse purchase.

Socrates says of the drunk man that he did not really *know* what he did was wrong, because he allowed himself to forget; he lost his focus and conviction. Aristotle would more simply say that the man suffers from *akrasia*, a weakness of the will. This is what Thunberg can't fathom about our mental handicap: how can we keep forgetting?! How can we be so *weak*?

The great philosopher of the Enlightenment, Immanuel Kant, most succinctly married knowledge with will in his formula: "Dare to know!" The intellectual virtues require the virtues of moral character, especially courage—knowing is an act of bravery. Enlightenment, Kant wrote in 1784, is our emergence from self-imposed immaturity ('nonage'). The cause of our immaturity "lies not in lack of understanding but in indecision and lack of courage to use one's own mind without another's guidance." The Enlightenment signifies our virtues—critical thinking and independent mindedness rather than blindly trusting in authorities or otherwise acting thoughtlessly. This is also the ethos of science.

But that's only half of the paradox! It's just the individualizing half. Remember, there is also the collectivizing half. Sure, science means thinking for yourself and questioning authority, but science also means expertise and authority. If the Centers for Disease Control and Prevention recommends wearing a mask to avoid spreading the coronavirus, are you really supposed to weigh that equally with your own Google research? Are we really supposed to set all the textbooks aside and think for ourselves, from scratch, about the composition of matter, space, and time? That cannot be—as Isaac Newton said a hundred years before Kant's essay: "If I have seen further, it is by standing on the shoulders of Giants." Science is a collective enterprise of social learning where each generation builds on the previous ones. So, to be scientific is both to distrust and trust. When to do one and when to do the other? There is no scientific answer for that.

As Latour (2004) argued, battlefield conditions have changed. For a long time, the right thing to do was to debunk claims to scientific authority, because they often were used to mask raw political power. Now, however, science, the media, the university, and other institutions of truth-making are in need of *bunking* and trust-building. This is especially true for climate change.

*   *   *

I suggested above that the ethics of energy is about complicity (see also Hughes 2017). Like any accomplice, you can point at the other guy and argue that your hands are clean. There is a strong argument to be made here that you are indeed innocent. In other words, individual agency is a moot point. The real culprit is the carbon-industrial complex. A 2017 Carbon Majors report, for example, determined that just one hundred companies were responsible for 71% of all greenhouse gas emissions (CDP 2017). Certainly what Latour (2018) calls the "obscurantist elite" and what Naomi Oreskes and Erik Conway (2010) call the "merchants of doubt" deserve more blame than others. The fascinating field of attribution science may be able to pin down legal (if not moral) responsibility with great precision. If a storm, for example, causes $1 billion·in damages, and it was made 20% stronger by climate change, and ExxonMobil has contributed a certain known percentage of carbon dioxide emissions, well, then we can derive a precise figure and send them the bill. This kind of moral logic is made stronger when a fossil fuel company knew about the climate dangers of their products but misled their shareholders and the public.

This carbon-industrial complex, run by men like the Koch brothers, buys politicians to push through deregulatory agendas. And they trick us by holding our attention on the benefits of fossil fuels while they smuggle all the hidden costs up their sleeves. A 2019 International Monetary Fund (IMF) working paper, for example, estimates that subsidies for fossil fuels in the United States reached $649 billion in 2015, far more than the budget for the Department of Defense (Coady et al. 2019). It may well cost us more to burn fossil fuels than to keep them in the ground. But that's not how things are made to *appear* to us. Poof! The rabbit disappears. What crisis, what war, what externalities? We don't *see* any of it. What we see is the new car in the garage and the frozen food aisle stocked full of ready-made dinners.

This brings up another paradox, one named for Enrico Fermi, the physicist who created the first nuclear reactor. Fermi's Paradox is roughly this: Given the very high probability that advanced extraterrestrial civilizations should exist, why have we not found any evidence of them? One answer is that civilizations keep reaching a stage similar to ours only to die off because they cannot devise a morality strong enough to contain their technological powers. Perhaps for us the reason for this is simply that big fossil fuel money too easily influences politics. The global elite have long known that the Titanic is sinking but they have been waging a campaign of doubt about climate change to keep the party going as long as possible. Short-sighted, selfish escapism may be the way this particular advanced civilization winks out of existence.

We can call this a campaign of *active* climate change denial (see Supran and Oreskes 2017). The superposition here is that this is simultaneously a matter of violating *and* upholding the spirit of science. As Oreskes and Conway (2013) argue, early modern scientists were vying for the social authority that had traditionally been conferred on the church. To win this battle, the scientists developed their own set of virtues. Most importantly, they placed a high burden of proof on their claims. They wanted a great deal of certainty about atoms or molecules or anything else before they would commit to believing in their existence. The philosopher William Clifford sums up this "Ethics of Belief" in his 1877 essay: "It is wrong always, everywhere, and for anyone to believe anything on insufficient evidence."

Most of all, modern science is set up to avoid "type 1" errors where a commitment is made to the reality of something that does not in fact exist. Let the church be embarrassed when its core metaphysical pillars (spirits, say, or creation in seven days) get knocked out from underneath it. Do not let science suffer such trust-eroding debacles! By contrast, "type 2" errors (not believing in something that does in fact exist) are considered more excusable. After all, there is time to gather more evidence to cautiously build the case for investing belief in something. Skepticism and doubt just mean withholding belief until sufficient evidence is gathered. What could be more rational, more scientific, more enlightened?

But what if time is running short? And what if the systems under examination are so complex, massive, and non-linear that the science keeps turning up uncertainties? Such is the case with the Earth's climate system. As the science policy scholar Daniel Sarewitz (2000) put it, paradoxically science often produces "an excess of objectivity." Every side of a controversial issue seems to have its own set of data to bolster its cause.

Despite its best efforts, science too is suffering the fate of the Catholic Church. The Protestant Reformation in science is upon us with what the transhumanist and social epistemologist Steve Fuller (2010) calls "Protscience" or Protestant science. Elsewhere, Fuller (2016) celebrates this as part of the Enlightenment project. He disparages the 'lions' like Latour who claim that truth must rest on the authority of shared practices, trusted institutions, and common cultural norms. By contrast, Fuller celebrates the 'foxes' who see all of this truth-making for what it is: acts of construction, or assemblages that can take different forms. He takes "dare to know!" to mean having the bravery to think outside the Catholic Church, if you will, of science (for Fuller, this especially means moving beyond Darwin).

The clearest example of Protscience is the Nongovernmental International Panel on Climate Change or NIPCC. The NIPCC is the product of the Heartland Institute, a right-wing think tank that makes hay pushing a thoroughly neoliberal agenda of deregulation and free market fundamentalism. What a Baptist church is to the Catholic

Church, the NIPCC is to the Intergovernmental Panel on Climate Change or IPCC. It keeps many of the trappings and serves the same social function, but it preaches a radically different doctrine. And, of course, it claims to be the real purveyor of the truth, or rather, it encourages its adherents to do their own research and not blindly trust institutions like the IPCC. In the age of balkanized social media where people are increasingly free to tailor their newsfeed, the NIPCC can preach to its flock without the interference of traditional gatekeepers.

I received a little book in my university mailbox from the NIPCC titled *Why Scientists Disagree about Global Warming* (Idso et al. 2016). They had mailed it to 350,000 high school and college educators. It was a condensed version of earlier, thousand-page reports produced by the NIPCC. I later attended a conference on US energy policy hosted by the Heartland Institute where the same little book was made freely available to the audience of prominent politicians, idea-makers, and industry leaders. The message at that conference and in the book is clear: They are not anti-science and they are *not* climate change deniers. No, they are actually the real scientists with the real science on their side.

The NIPCC book never specifies what 'disagree' means in the title about why scientists disagree about global warming. This is the key epistemic move. Imagine all the variables involved in climate change: extreme weather, soils, the stratosphere, the role of water vapor and clouds, the human contribution, the placement of rain gauges and thermometers, ice cores, polar bear populations, the right policy response, the proper estimation of uncertainties, the right inferences from this or that data set, and so on. No scientist is going to agree with another scientist across the board on all of this. Thus, scientists disagree.

This is a page taken from the early modern philosopher Thomas Hobbes who recognized that the empirical sciences could never fulfill their promise to transcend all religious and political factions to deliver the one true picture of nature. They could not legitimately referee and resolve political debates, because they would never be able to agree among themselves (see Shapin and Schaffer 1985). As Hobbes said, we will continue to "see double" and multiple truths will legitimate multiple claims to authority and multiple policy agendas. The NIPCC and the IPCC are pursuing incommensurable research and policy agendas all under the banner of science. The result is the same old religious battles only now gussied up as science.

Of course, just because one group says A and another group says not-A, doesn't mean there is no truth of the matter. To chalk this up to a he-said/she-said disagreement is exactly what the NIPCC wants. As Kant noted, part of Enlightenment means being able to discern real experts from charlatans. NIPCC is devious, irresponsible, and crazy. Their report is a comedy of contradictions. For example, it notes that "true science is never settled." It is always provisional and open to falsification. Disagreement in science, then, is "the rule" and that is because "science is a process leading to ever-greater certainty, necessarily implying that what is accepted as true today will likely not be accepted as true tomorrow" (p. 9). But what does this mean?! We used to think illness was an imbalance in the humors and that fire came from phlogiston. We *think* we know better now, but give it time. Bacteria, oxygen and the whole periodic table of elements, plate tectonics—these are "likely" destined for the dustbin too. Don't believe anything until you have sufficient evidence—and we *never* have sufficient evidence.

The image they paint is that science is a process, a fluid, and thus not suitable as a foundation for policy. Yet the NIPCC book inconsistently insists that the "formulation of effective public environmental policy must be rooted in evidence-based science" (p. 83). So, science is never settled and can't be trusted but we have to trust science. This is how the obscurantist elite do their little sleight of hand card tricks. They weaponize science's core virtue of avoiding type 1 errors by declaring that we must have the best available science. We must keep raising the bar for quality evidence so that we can be certain before we act. That is, after all, the only rational and ethical way to behave! But it amounts to the collective irrationality and immorality of partying on the Titanic. The ship is sinking to the soothing Muzak of science, the soft jazz of facts, which like any good piece of jazz can be interpreted in a number of ways.

The NIPCC is an example of the shadow cast by the Enlightenment. In their book, they note that the complexity of climate makes it "difficult for unprejudiced lay persons to judge for themselves where the truth actually lies in the global warming debate." Therefore, these lay persons turn to the "supposedly authoritative statements issued by one side or another in the public discussion." But "Arguing from authority…is the antithesis of the scientific method" (p. 59). As Kant said, Enlightenment means daring to know and thinking for yourself. To blindly trust the IPCC is the same kind of immature behavior that held people back in the Dark Age.

Of course, the NIPCC book saws off the branch it is sitting on: if you can't trust authoritative statements on *either side*, then that includes them. But this doesn't matter for their strategy. They only want doubt. They don't want belief; they want its suspension. The method is in some sense Socratic, because it is a negative dialectic, which questions all claims to truth. Socrates often talked his interlocutors into a state of *aporia* or impasse and puzzlement. They were paralyzed, which is why Socrates was known by some as the sting ray more than the gad fly. But in an industrial society premised on growth, the result of a negative dialectics is not paralysis; it's the continuation of the status quo, namely drill and burn and profit as the climate inches closer toward the tipping point.

*   *   *

These *active* deniers, these sophists, who profiteer by sowing doubt, buying politicians, and performing sleight of hand tricks with economic accounting and pseudo-scientific positions have truly sunk into vicious greed. In particular, the Republican Party of the United States needs to be called out as a dangerous, irresponsible organization for its active denial of climate change. Again, sometimes when we encounter A and not-A (IPCC and NIPCC) we are not dealing with a paradox; we are dealing with reliable information versus disinformation. Daring to know entails overcoming the laziness of false equivalencies and the urge to confirm your biases by believing just what you always wanted to believe anyway.

But can we lay *all* the blame on some corporations and politicians without it trespassing into our lives? Could we excise those hundred most-polluting companies without pulling on a hundred million strings? How are we going to transition to a carbon-free energy system without negotiating our carbon-based way of life? Because we can't imagine anything but a smooth transition where life just keeps humming along, we become the new "silent majority," the *passive* deniers. Let's face it: we want to be fooled! We don't want to see what's up their sleeves, let alone understand how the magic trick works. We certainly don't want to pay the real cost for a gallon of gas. We want the reassurance that everything is *under control*. We don't want to think CARBON in the morning; we want to think BACON (Franzen 2019).

All this energy! So much is done yet nobody seems to be responsible for doing anything. That Nobody is full of energy, but we are enervated. Like energy itself, the ethics of energy is neither here nor there. It's not something we can locate. It's something that *just happens*. All this activity, but our ethics comes in the *passive* voice.

This passive denial is a form of immaturity in the way that Kant described it—letting others do and think for you. It is a denial not just of our responsibility but also our solidarity. Remember that thinking CARBON in the morning is an acknowledgment of what Schrödinger called *Verschränkung*, entanglement. My life is tangled up with yours and human lives are tangled up with non-human lives. The problem with an ethics of complicity is that it lets everyone off the hook. If individual agency really is moot, then each individual can think: "If I didn't do any of this, then I am not involved. Don't drag *me* into it!" But you have already been dragged into it! You are just in denial, the coping mechanism of those who suffer from *akrasia*, the mental illness of the perfectly sane and normal people walking around in only half of reality.

Insulation is not just a material for energy efficiency. It is our basic way of being in the world. We wrap ourselves in layers and layers of moral insulation. Unfeeling, we rampage the Earth. We are numb. If only Volta could give us a little zap from his electrifying apparatus.

To take Greta and the other young climate strikers seriously would be almost impossible. What they are suggesting is that we are all quite calmly participating in a system of death and annihilation. Arendt coined the phrase "the banality of evil" to explain how thousands of Nazi functionaries participated in mass murder only to come home at the end of the day to share a nice steak with the wife and kids. Similarly, those of us in rich countries spend our days consuming fossil fuels like crazy only to come home and rest peacefully at night having disassociated ourselves from the climate crisis and the sixth mass extinction. We are good, decent people. But are we *also evil*?

What enabled the Nazi functionaries to go about their business, Arendt concluded, was their inability or their refusal *to think*. They just built walls around their little-old-selves to see only one half of their superposition. When Eichmann, who kept the trains running on time to the concentration camps, was asked how he could commit murder, he replied: "Ich habe niemals einen Juden getötet." Roughly, "I never killed any Jews." He just pushed the buttons. He just followed orders. He was an individual. He was not the system.

Our actions are also mediated through an impossibly complex system. "I didn't cause climate change. I just drove my kids to school. I just took a harmless vacation to Europe." Between doer and consequence are ample mitigating circumstances. You can draw the line between self and non-self in so many ways, which means we can't pin any crime on anyone. Again Heraclitus was right: all is flux!

I said above that we feared obsolescence when it comes to work, but we have already embraced our obsolescence when it comes to morality. The individual, that traditional unit of ethics, is antediluvian. Just sit back and enjoy the show. Savor the hamburger, because depriving yourself will not make one jot of difference. I think about this when I bicycle to work. As if me foregoing one car trip makes a dent in a world where over 100,000 planes fly every day! It's like whispering sweet nothings into a hurricane.

But I bicycle anyways, because it is a way of putting skin in the game. To feel the hills in my legs rather than to have them hidden under the gas pedal is a way to *feel the reality* of our situation. I try hard also to eat less meat, as a way to remember daily our solidarity and not have it slip my field of mental attention. The moral imagination we are trying to expand requires us to fight against the forgetfulness a high-energy life encourages.

Hans Jonas (1984) put it this way: we need to expand our ethics to match the expanded scope of our technological powers. But how do we make those powers a present reality, a part of our moral consciousness? His solution: we need a new kind of fear. Presently, fear is an emotion reserved for our snake brain. It is the flight-or-fight jolt we cannot help but feel. Alas, though the climate crisis is fast moving on a geological scale, it is too slow moving and diffuse to zap us with this kind of fear. We automatically fear weather (a tornado, say), but we have to *learn* how to fear climate (perhaps in the same way we learn to fear the wrath of God). We need to cultivate a higher-order kind of fear, a cerebral scare. The climate strikers are more mature than their parents' generation, because they have developed this new skill, this new *virtue*.

The orthodoxy insists on rationality and pooh-poohs fear. Don't be fear-mongers! Yet the only kind of rationality it knows is coldly instrumental. It is so clever about the means and clueless about the ends. Our "way of life" is sacrosanct. It's as sturdy as the burger joint we just built but already can't remember what life was like before it and can't imagine life after it. We need to learn the rationality of fear and the prudence of extremism. This is the lesson that David Wallace-Wells (2019) is helping a new generation of scientists understand. I often ask my ecology students if perhaps their job is to be false prophets. Maybe they need to warn us of a world that never becomes a reality because they scare us out of it.

The active deniers have adopted the amorality of Nietzsche's master race, those "exuberant monsters that quit a horrible scene of murder, arson, rape and torture with the high humor and equanimity appropriate to a student prank" (in Mason 2019). What Thunberg can't imagine is that this 'equanimity' also characterizes so many of the *passive* deniers too. We say we know that we must act, but we don't upend our lives. We are so calm in the midst of frenzy. It is so real that it feels un-real, like a prank. Everything will be alright in the end. We just know it. Actually, we really *believe* it despite everything we know.

These monstrous times! Just when the stitching and stapling of a high-energy global economy is pulling us all more tightly together, we are also tearing asunder. The energy journalist David Roberts (2017) puts this in terms of "tribal epistemologies," or incommensurable views about truth. One planet, many worlds. We just don't *believe* each other. It is little wonder that empathy has declined as the high-energy society has wrapped us all in our own worlds (Szalavitz 2010). The ultimate superposition: we inhabit alternate realities at the same time in the same place. Even before the rise of deep fake videos, we have learned that seeing is *not* believing.

Imagine if Dr. Frankenstein visited the scene where his monster had committed murder and instead of being wracked by guilt and righteous indignation (those hot Victorian virtues) he just smirked and said "prove it" (like the active denier). Or imagine he confessed and said "I know, what a shame!" but then *did nothing* to right the wrong (like the passive denier). What if he didn't *dare* to know? No amount of science, no bulldozer of proof, could budge a man so coldly devoid of virtue. The climate strikers are alerting us to our own heat death. No amount of voltage could zap such a heart back into sinus rhythm. Courage is the human prime mover. What the climate strikers are saying is that we are cowards, and that *this* is our "energy crisis."

Bibliography

CDP. 2017. The Carbon Majors Database: CDP Carbon Majors Report 2017, July. https://6fefcbb86e61af1b2fc4-c70d8ead6ced550b4d987d7c03fcdd1d. ssl.cf3.rackcdn.com/cms/reports/documents/000/002/327/original/Carbon-Majors-Report-2017.pdf?1501833772

Clifford, William. 1877 [1999]. The Ethics of Belief. In *The ethics of belief and other essays*, ed. T. Madigan. Amherst: Prometheus, pp. 70–96.

Coady, David, et al. 2019. IMF Working Paper. Global Fossil Fuel Subsidies Remain Large: An Update Based on Country-Level Estimates. *International Monetary Fund*. https://www.imf.org/en/Publications/WP/Issues/2019/05/02/Global-Fossil-Fuel-Subsidies-Remain-Large-An-Update-Based-on-Country-Level-Estimates-46509

Copleston, Frederick. 1993. *A History of Philosophy, Vol. 1, Greece and Rome from the Presocratics to Plotinus*. New York: Doubleday.

Dupuy, Jean-Pierre. 2009. The Precautionary Principle and Enlightened Doomsaying: Rational Choice Before the Apocalypse. *Interdisciplinary Studies in the Humanities* 1 (1), October 15. http://occasion.stanford.edu/node/28

Franzen, Jonathan. 2019. What if We Stopped Pretending. *New Yorker*, September 8. https://www.newyorker.com/culture/cultural-comment/what-if-we-stopped-pretending

Fuller, Steve. 2010. Science in God's Image. *The Guardian*, May 3. https://www.theguardian.com/commentisfree/belief/2010/may/03/science-religion-intel-ligent-design

———. 2016. Science Has Always Been a Bit 'Post-Truth.' *The Guardian*, December 15. https://www.theguardian.com/science/political-science/2016/dec/15/science-has-always-been-a-bit-post-truth

Hughes, David McDermott. 2017. *Energy Without Conscience: Oil, Climate Change, and Complicity*. Durham: Duke University Press.

Idso, Craig, Robert Carter, and S. Fred Singer. 2016. *Why Scientists Disagree About Global Warming: The NIPCC Report on Scientific Consensus*. Arlington Heights: The Heartland Institute.

Jonas, Hans. 1984. *The Imperative of Responsibility: In Search of an Ethics for the Technological Age*. Chicago: University of Chicago Press.

Kant, Immanuel. 1784. What Is Enlightenment? Mary C. Smith, trans. http://www.columbia.edu/acis/ets/CCREAD/etscc/kant.html

Latour, Bruno. 2004. Why Has Critique Run out of Steam? From Matters of Fact to Matters of Concern. *Critical Inquiry* 30: 225–248.

———. 2018. *Down to Earth: Politics in the New Climatic Regime*. Medford: Polity Press.

Mason, Paul. 2019. Reading Arendt Is Not Enough. *The New York Review of Books*, May 2. https://www.nybooks.com/daily/2019/05/02/reading-arendt-is-not-enough/

Oreskes, Naomi, and Erik Conway. 2010. *Merchants of Doubt: How a Handful of Scientists Obscured the Truth on Issues from Tobacco Smoke to Global Warming*. New York: Bloomsbury.

———. 2013. The Collapse of Western Civilization: A View from the Future. *Daedalus* 142 (1): 40–58.

Roberts, David. 2017. Donald Trump and the Rise of Tribal Epistemology. *Vox*, May 19. https://www.vox.com/policy-and-politics/2017/3/22/14762030/donald-trump-tribal-epistemology

Sarewitz, Daniel. 2000. Science and Environmental Policy: An Excess of Objectivity. In *Earth Matters: The Earth Sciences, Philosophy, and the Claims of Community*, eds. Robert Frodeman and Victor R. Baker. Upper Saddle River, NJ: Prentice-Hall. pp. 79–98.

Sarewitz, Daniel. 2004. How Science Makes Environmental Controversies Worse. *Environmental Science & Policy* 7 (5): 385–403.

Shapin, Steven, and Simon Schaffer. 1985. *Leviathan and the Air-Pump: Hobbes, Boyle, and the Experimental Life*. Princeton: Princeton University Press.

Supran, Geoffrey, and Naomi Oreskes. 2017. Assessing ExxonMobil's Climate Change Communications (1977–2014). *Environmental Research Letters* 12 (8): 1–18.

Szalavitz, Maia. 2010. *Born for Love: Why Empathy Is Essential – And Endangered*. New York: Harper.

Thunberg, Greta. 2019. School Strike for Climate: Save the World by Changing the Rules. *Ted*. Transcript at https://www.ted.com/talks/greta_thunberg_school_strike_for_climate_save_the_world_by_changing_the_rules/transcript?language=en

Wallace-Wells, David. 2019. *The Uninhabitable Earth: Life After Warming*. New York: Tim Duggan Books.

Part II

Energy Orthodox

Had God intended the work of the world should be done by human bones and sinews, he would have given us an arm as solid and strong as the shaft of a steam engine.
Horace Mann (1853)

7

First World Problems

Nature, to be commanded, must be obeyed
Francis Bacon, *1620*

Dear reader, I have had an energy transition and converted to the orthodoxy. Actually, I don't really like that term 'orthodox.' It sounds so rigid and narrow. In fact, we pitch a big tent with plenty of room for different perspectives. We house, for example, the crucial debates about nuclear power—is it the key to solving climate change or not (e.g., Shellenberger 2017 vs. Jaczko 2019)? If such significant differences occur *within* the orthodoxy, what's the point of talking about the orthodoxy at all? This is to paint with such a broad brush as to create a night when all cows are black! It is to erase all the significant distinctions.

Let me put the point more generally. There are three approaches to climate change: mitigation (reducing greenhouse gas emissions), geoengineering (direct interventions in the climate system), and adaptation (adjusting to the effects of climate change). Drawing from David Keith (2007), we can represent these three strategies with a simple diagram. The

© The Author(s) 2021
A. Briggle, *Thinking Through Climate Change*, Palgrave Studies in the Future of
Humanity and its Successors, https://doi.org/10.1007/978-3-030-53587-2_7

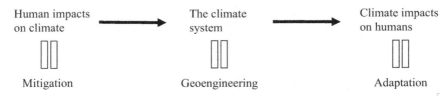

Fig. 7.1 The three strategies for climate change policy

point is that any *serious* proposal for any of the three strategies will come from *within* the orthodoxy (Fig. 7.1).

Earlier, I doubted the instrumental rationality of the energy orthodoxy. It's true that major institutions and policymakers focus almost entirely on instruments, but now I see that they do so for a couple of good reasons. First, we really do know what the ends are. We know progress when we see it. Imagine a poor child living in an efficiency apartment in Maryland where cockroaches and mice scurry across the floor. His family barely has enough money to keep the lights and heat on and food in the pantry. If that child grows into a man whose own children live in a nice, clean house with ample food and opportunities, even family vacations, well, we know he has moved up the ladder. It is self-evident.

Second, this means that all the important ethical questions are instrumental—they are about *how to* get people up that ladder in ways that are just and sustainable. Whether to commit to nuclear power, how to store solar-generated electricity, how to distribute wealth and responsibility, how to fashion a more sustainable food production system, and how to design more efficient homes and cities are some of the most pressing ethical questions facing civilization. On the grandest scale: How do we decarbonize the economy? Getting the right mix of tools is *the* moral responsibility of our age.

If I can confess something to you, I never was comfortable with all this talk of 'virtues.' Doesn't it strike you as elitist? Who are we to say what is high and low, noble and base, proper or inappropriate? I recall that Epictetus quote about freedom coming not from having the things we desire but by "removing the desire." That's crazy. Do we really expect people to "remove the desires" they feel? Is that fair? Is that a viable political project?! And it doesn't matter anyway. For example, if every American cut their meat consumption by 25% (I have no idea how you would get 330 million people to do that!), it would prevent 82 million metric tons of greenhouse gasses annually (Mock and Schwartz 2019). That sounds impressive, but it is the equivalent of just *two days* of emissions from China. We need to work on more 'virtuous' systems, which just means better technology and smarter policies. The 'virtues' (to use that term tongue-in-cheek) that matter are those of the engineer, scientist, and entrepreneur.

Moral virtues operate by the doctrine of the mean—not too little, not too much. That sounds nice, but it is question begging. Of course we should avoid excesses. But how is "too much" going to be defined? By what standards? More importantly, by *whose* standards? Who will decide what counts as excess energy consumption? Are they going to have the authority to compel others to abide by their definition of the mean? What I have come to see is that anything more than the orthodoxy's focus on better tools and growing the economic pie becomes a slippery slope to draconian regulations. Any proposal to limit energy consumption will drag us into interminable moral disagreements and intolerable political coercion. In short, it is good that we live in the age after virtue. It is good that the meaning of 'good' is left up to individuals.

The so-called orthodoxy is even big enough to encompass the paradoxes or superpositions noted above. We acknowledge that these often indicate real problems. It's just that these are first world problems.

By definition, first world problems are less problematic than the alternative. The developed nations that create these problems also create their solutions. Indeed, development is the gradual shrinking of problems by making each round of problems less problematic than the ones we had before. In the seventeenth century, 'teeth' was listed as the fifth leading cause of death. To be sure, our modern sugary diets give us cavities and we need to do better at ensuring equitable access to dentistry, but these are easier problems to solve than the ones we had before modern dentistry (see Otto 2017). Occasionally, my washing machine clogs up and I have to tinker with it. But better this than filling up the old wash bin or hauling laundry to the river to spend hours of back-breaking labor scrubbing away.

Perhaps there are excesses in the developed world. In the scheme of things, though, that doesn't matter. As noted in the *BP World Energy Outlook* 2019, "all of the growth in energy demand" through 2040 will come from developing economies. Billions of people still have far too little energy, which makes any talk about limits or virtues irrelevant at best. At worst, it is immoral to focus on limits when so many people are poor. The defining ethical question we face is instrumental: How to provide the energy for development without destabilizing the climate system?

Right now, the energy developments around the globe are like the fiery boosters on a rocket ship. We are accelerating and, admittedly, we are burning lots of fuel and charring huge chunks of Earth on our way toward escape velocity (Karlsson 2015). But we cannot let the paradoxes paralyze us! It may be tempting to ease up, but in fact we must pull even harder on the throttle if we are to *breakthrough*. We will get to the Good Anthropocene, that Promised Land of technoscience, that heaven of our *own* divine making where there will be no more paradoxes or tragedies. Driving your car will actually clean the air. Laundry is another good example, where modern machines are increasingly energy and water efficient. Yes, we are creating monsters, but the real moral failing is in the abandoning, not the making. We must "love our monsters," not run in fear from them (Shellenberger and Nordhaus 2011). We just need to make our machines behave better and better—this is the (dare I say) *noble* project of "instrumental rationality."

What, you may ask, about those supposedly dark, upside-down energies? We long ago stopped believing in such fairy tales. There is no megamachine. There is no capital T Technology from Heidegger's *Gestell*. There are only all our little babies, those little t technologies of tinkering. This is not really about modernity so much as it is the age-old, universal human quest to control the environment around us. No one has ever yearned to wake up and spend the day scrubbing clothes against the river rocks! The Anthropocene is the story of all humanity; it is the story of our shared need to build a home on Earth. Ours is a pantheistic religion. Forget the gods of forest and field, we worship our sacred technologies!

My apologies. I need to take a breath. In my newfound zeal, I have gotten ahead of myself. I risk scaring you away. I don't need to proselytize, because you already believe everything I am going to tell you. You just don't really know that yet. Let me start with a parable.

Imagine a sunny day in Puerto Quetzal. The cruise ship is docked and the tourists are taking selfies. A volcano in the distance looks like the profile of the head of an enormous fish-god with its rocky lips jutting skyward. Exotic birds soar above the green foliage and beyond the blue waters. The sea air refreshes the lungs and rejuvenates the spirit. One of the tourists, a businessman from America, notices something in the frame of his picture. There, in the foreground on the shore, is a small fishing boat with a poorly dressed local fisherman dozing on the bow.

Feeling helpful, the tourist walks over and asks, "What are you doing?"

In decent English, the fisherman says, "I have just come back from the morning catch."

"There is plenty of daylight left, will you head back out to catch more?"

"No," the fisherman replies, peering up from under the brim of his hat, "I have caught enough for the next couple of days. I am going to rest here a while and then play with my kids and go dancing with my lady."

"But if you go back out you can catch more, right?"

"Yes, I suppose…"

"With the extra income, you could buy a motor for your boat and then catch even more fish each day."

"Yes, and then?" the fisherman asked.

Now the businessman was feeling very helpful indeed. A little pro-bono advice for the locals. "Well, then, you save up for another boat."

"I am only one man, how could I sail two boats?"

"Ah," now the businessman was excited, "that's the thing! You'll have enough income to hire someone to help you."

"Oh…"

"And then your catch will grow. Eventually, you can invest in a cold storage plant or a pickling factory. You could ship your product directly to Miami and Chicago. You'd be able to move to a bigger city to run your business. It would be a lot of work, but then…"

"Yes, then what?" the fisherman wanted to know.

"Well, then you'll be free to take vacations where you could doze by the side of the sea, play with your kids, and go dancing with your lady!"

"Ah," the fisherman said, lowering his hat back over his eyes, "that sounds like a good life."

*   *   *

This parable has been told many times. It originates from a short story, penned in 1963 by the German writer Heinrich Böll, titled "Anecdote concerning the Lowering of Productivity" (Böll 1986). Böll's story features a tourist who confronts a 'lazy' fisherman wasting time at an unnamed harbor somewhere on the west coast of Europe. In his story, the businessman leaves the encounter flustered and envious. For Böll, the fisherman is the wise one. But he is wrong. The energy orthodoxy draws its power and moral clarity from the *right* interpretation of this parable. The businessman is the wiser of the two.

To understand this, let's build back up to the story from the primal energy transition of life: metabolism. From single-celled organisms to whales, all living creatures must continuously make and re-make their bodies from external parts. This is a radical form of vulnerability; even a brief pause in the energetics of this ceaseless self-building leads to death. Life is risk. The ethics of energy, then, is rooted in what we might nowadays call risk assessment and mitigation.

One strategy to decrease vulnerability is to build armor around the fragile self. Insects with exoskeletons do this as a physical part of their body schema. Other animals build shelters to aid in their survival: beavers build dams, birds build nests, bees make hives, and so on. Humans do this world-building with unprecedented intensity and complexity. To be human is to be born not just on the Earth but into a world that is the product of human work. For humans, energy truly is the capacity to do work because we use energy to make a home through intelligent work. This is so even if it is as simple as being swaddled in the fur of animals, sitting around the fire, and wearing a bead necklace. This world is our culture or "second nature."

It begins with our upright posture. The bipedalism of human walking costs about 75% less energy than both quadrupedal and bipedal walking in chimpanzees. This initiates a cascade of evolutionary moves. The arms are freed up to carry items, allowing for nomadic forms of mobility with roving home bases. Bipedalism creates a more plastic coupling of respiratory rhythm and gait, permitting more complex vocalizations. The speaking animal forms intricate social arrangements and conveys rich and abstract meanings. The arms are also free to make and use tools. The speaking, tool-using animal develops a complex brain. The brains of other primates consume 8–10% of resting metabolic energy. The brains of *Homo sapiens* consume 20–25% (Smil 2017).

That is our energetic gamble: strong brain, weak body. (It always comes down to energy trade-offs: the first law of thermodynamics tells us there is no free lunch.) The brain allows us to appropriate exo-somatic forms of energy in ways other than simply ingesting calories. Our metabolism, in other words, spills outward and we build the world as an extended body just like we build our biological bodies. We master fire for heat, light, safety, cooking, smelting ores, and even shaping entire ecosystems. We develop simple tools to gain mechanical advantages for hunting, building, artistry, and farming. Then we develop complex machines. Our machines become bodily extensions—as Horace Mann noted, we build steam engine arms. This is as God (aka evolution) intended.

(To risk getting ahead of ourselves again, eventually, we will merge thoroughly with the machines to become super-intelligent beings. As James Lovelock (2019) predicts, the Novocene, the age of presently unimaginable intelligence, will soon follow the Anthropocene.)

As it develops, culture reduces our dependence, not so much on nature per se (as even the most technological world is derived from nature and, in some sense, is itself natural), but on fate or *fortuna*. We decrease our exposure to the vicissitudes of this or that particular place at any given moment in time. Risk is reduced. In a word, we gain control. In the Greek myths, the Fates symbolize those natural and supernatural forces that bring forth, measure out, and apportion (perhaps they are the dark side of virtue, which is also about measure and proportion). They give to each their portion or share that defines the appropriate limits. The Fates have many genealogies, but are in one way or another the children of death, of the order of things, or of mother Necessity. They must be propitiated and, insofar as possible, their power must be neutralized.

All cultures have sought to shield themselves from the Fates with varying degrees of success. They have used rituals, incantations, superstitions, and offerings. Their augurs have interpreted omens from the cracks in a turtle shell to the alignment of stars. Their shamans have danced with spirits and energies of all sorts. Most cultures have not been very successful. The Black Death killed as many as 50% of all Europeans in the fourteenth century. The Catholic Church was useless and priests died along with everyone else. Even into the mid-twentieth century the Third Pandemic of bubonic plague killed millions of people in China and India.

Our culture has hit upon the winning formula and what we call progress or development now is the uptake of this recipe across the globe. Get the plague today? Take an antibiotic pill! A simple word for this formula is science, which is really just a radical precisioning of the universal human ability to notice patterns amidst change (to see the self-same E behind nature's dance of the seven veils). We note, for example, that the seasons bring growing cycles, allowing us to plan and predict the migrations of prey animals and the growth of crops. We notice the fleas on the rats and the bacteria on the fleas. We increasingly trace the natural connections rather than imagine the supernatural ones.

Another word for this formula is Enlightenment, and as its contemporary defenders note, it is doing an impressive job of replacing really bad problems with more manageable ones (see Pinker 2018; Rosling 2018). Extreme poverty, child mortality, and hunger are plummeting. Life expectancy, literacy, and education are on the rise (see Matthews 2018). Most of these improvements are made possible by our mastery of dense sources of energy, especially fossil fuels. Wind, solar, and battery prices are all plummeting, meaning that we are learning how to master other forms of energy that can keep our improvements rolling with fewer unintended harms like climate change.

The virtues of this way of life are a complex web of reason and humanism. At the center of this web is Francis Bacon's reality principle that to control nature we must obey her. If we discipline our gaze strictly on natural phenomena, gradually we eliminate superstitions and supernatural bogeymen. The spirits animating the natural world come into focus as, for example, ATP, the molecular currency of intracellular energy transfers. The ghosts in the lightning storm are unmasked as electrons. The electrons are understood as part of a field involving magnetism. The laws governing such fields are quantified and put to work in generators and other machines. Ply the screws to nature (following her own rules!) and she spills her secrets. It may be a nasty business, but remember she is the one who terminates most lives with a final act of gruesome murder. She is the mother of plague; we are the ones on the humanitarian mission!

The mission is about control in the name of reducing risks or exposure to the Fates. We go from helplessness in the face of diabetes to learning how to get insulin from pigs to learning how to generate it synthetically in labs by genetically rewriting bacteria to produce a human hormone. This is a story of increasing control and liberation. People benefiting from Enlightenment are free not just from bubonic plague but many other diseases that once decimated whole populations. Anti-vaccination movements that cause measles outbreaks demonstrate how hysteria can lead to a regression. When we lose our nerve or our faith in science, we can slide back from first world problems to the worse kind.

I highlight medical developments intentionally to indicate the breadth of the energy orthodoxy. Without mastery of energy sufficient to make it cheap, plentiful, and reliable, there wouldn't be the social stability or the material infrastructure needed to sustain the laboratories where the delicate work of dismantling the Fates is conducted. No E, no vaccine for the coronavirus, and certainly no mass production system to get it to billions of people. It admittedly takes a great deal of force and grime to produce sterile equipment. If you doubt whether the costs are worth the benefits, just wait until the next time you or your loved one needs to visit the emergency room.

* * *

This brings us back to our parable. The fisherman may be set for a couple of days, but then what? He is resting in the fickle hands of fate. What if a storm blows in, his wife falls ill, or his boat is damaged? There is no fat stored up, no reserves to *ensure* the bottom won't simply fall out tomorrow. As such, the fisherman cannot really be resting at ease. He is trapped on the wheel of necessity, forced to continuously wrest his existence from fickle and powerful forces. In energetic terms, there is not much of a surplus. The fisherman's life is as lean as the fish studied by the ecologist Charles Hall (2011) who invented the concept of energy return on investment or EROI: how much energy is returned from one unit of energy invested.

By contrast, the businessman has a savings account, insurance, and a retirement plan. He's got ample food in the pantry and more than he can shake a stick at down the street at the grocery store. He has stored up a high-energy profit. As a result, he is truly carefree. He is like the modern camper ensconced in his high-tech tent with canned food and all sorts of gadgets at the ready. Now he can relax, look up, and really enjoy the stars!

The point is much larger than an individual's financial and food security. The modern world represented by the businessman is a network of insurance—a grand risk-reduction scheme. The electricity grid supplies reliable power for hospitals as well as the scientific labs where the medicines are developed. Buildings provide climate-controlled environments to sustain life against the onslaughts of extreme cold and heat. Fossil-fueled machines both produce greater agricultural yields and stitch the world together so that a poor crop in one place is balanced out by a bumper crop elsewhere. Systems of reservoirs, purification plants, pumps, and pipes deliver reliable drinking water. Information networks allow for rapid diffusion of knowledge to solve problems.

Energy is fundamentally about control and control is about security. This is a point that Daniel Yergin, the world's leading energy analyst and founder of CERAWeek, consistently underlines, including in his magisterial history of the geopolitics of energy *The Quest: Energy, Security, and the Remaking of the Modern World* (2011). Reading that book as well as Yergin's Pulitzer Prize history of oil (Yergin 2008) makes plain the central importance of securing access to energy resources.

Those of us in the developed world have forgotten just how much security gas lamps and then electric lights brought to streets that were once plunged in nighttime darkness. In 1816, the American entrepreneur Rembrandt Peale had a "magic ring" of one hundred burners at his gasworks. He used a control valve to turn the ring of flame up and down to the crowd's amazement. Imagine an entire population wowed by a dimmer switch.

The fisherman suffers from energy poverty, a lack of access to modern energy services. Admittedly, providing those services brings new problems but they are far better than what millions of people today must deal with. Various definitions for "energy poverty" exist (see Birol 2007; Lavelle 2013). Yet definitions are less important than imaginations. Imagine, for example, your life without electricity, which is the reality for 1.6 billion people, predominantly in Africa and south Asia. This means no safe, reliable lighting, no refrigeration for food and medicines, and no air conditioning. Or imagine cooking your food or heating your home with biomass such as gathered twigs and dung, which is the reality for hundreds of millions of people worldwide. Imagine the time spent simply collecting the fuel needed for the day as well as the resulting deforestation. The hardships and injuries that result from this labor often fall disproportionately on women and children (Gaye 2007). Annually, exposure to hazardous air pollution from inefficient stoves kills hundreds of thousands of people (Chafe et al. 2014).

Now, our successes in obtaining reliable, dense energy stores has created the new problem of climate change. Yet climate change, too, is a first world problem. That is not to dismiss it, just to put it in perspective. Addressing climate change requires carefully considering other problems associated with energy systems, many of which may be more severe. Renewable energy mandates, for example, can increase energy costs in ways that act like a regressive tax harming the most vulnerable (Lomborg 2014). As another perspective-check, consider that climate-related deaths actually decrease sharply as development intensifies (see Epstein 2014). Furthermore, weather-related disaster losses from 1990 to 2016 as a proportion of overall GDP have *decreased*. Indeed, the decrease in losses as a fraction of GDP are greatest in the poorest countries (Watts et al. 2019). That's because with modern energy services or 'volts,' we literally build shelters from the storms.

Climate change is more like "diabetes for the planet" than an imminent asteroid (Nordhaus and Trembath 2019). Managing it requires lots of policy moves and plenty of R&D, but it does not require draconian limitations on freedom or consumption. Indeed, the worst-case climate future would be one where people don't have the wealth to afford new technologies and the adaptive resilience they provide. That's why climate change should be framed in terms of "human dignity" rather than "human sinfulness" (Prins et al. 2010). We are creating new climate risks with fossil fuels, but fossil fuels have built a world that supplies the wealth and innovation necessary to manage the new risks. High-energy civilization shields us from the risks of a climate that has always been deadly. We may be exacerbating floods and drought, but we didn't invent them! As the song goes: "we didn't start the fire."

In sum, the modern world achieves unprecedented resilience and security by neutralizing the Fates and their minions: famine, flood, drought, and disease. The moral of the story for the energy orthodoxy is that it is best to develop the infrastructure capable of delivering a life of comfort with a backstop of security. If the businessman is jealous of the fisherman, it's only because that backstop is working so well that it has become invisible. Or the (first world) problems posed by that backstop (e.g., long wait times to speak with a human agent at the credit card company) are not put in proper perspective.

Perspective is often lost, because the media keep exaggerating problems to get clicks. The negativity bias of the media is extremely important. The truly momentous good news about the human condition that has been made possible by high energy often unfolds across years or decades, which is far too long to capture the attention of the media. By contrast, disasters and other forms of bad news tend to strike instantly. The overall impression is that the world is getting worse when the reality is the opposite (see Lomborg 2001).

The global fear and outrage about the Amazonian rainforest fires in Brazil in 2019 is a good case in point. Celebrities and even world-leading politicians warned that the "lungs of the earth" were on fire, threatening global oxygen supplies. Such stories gave the impression that the Amazon was on the verge of disappearing. Yet in reality, that fire season was not all that remarkable, oxygen supplies are not threatened, and Brazil has managed to conserve 80% of its portion of the Amazon compared with pre-1970 levels (Shellenberger 2019). Many of the celebrities' pictures were not even real. This isn't to deny real problems. It's just to point out how easily media-fueled hype can displace the level-headed, pragmatic thinking, which is essential to the orthodoxy's ongoing success. In this case, the solutions lie in working with local populations to ensure they have enough security and wealth to care for the rainforest. We need to put incentives in place to make conservation profitable.

For a variety of reasons, then, the businessman might forget momentarily that his vacation has only been made possible by the enormous network of consumption and production. He might lose sight of the bigger picture and feel less than satisfied or even anxious about his way of life. Yet if he were to actually change places with the fisherman, he would quickly realize all the assurances he had taken for granted. He'd come running back to the shelter of his home.

By commanding ever-greater stores of energy, we can control our own fate and conjure what we need on demand without having to subjugate ourselves to the limits of place or season. After all, the fisherman had a boat. He didn't rest content with his bare hands and other limits of his own body. The boat extends his bodily powers, thus allowing for a more reliable and productive catch. Why not, then, put a motor on the boat? Why not a fleet of boats? The excess fish captured will spoil if stored too long for personal use, but if sold, the money becomes the capital needed to build the world that reduces vulnerability. Eventually, we can have robotic self-operating boats or even 'fish' grown in a lab. We can enjoy a guaranteed 'catch' while lounging on the shores of a sea teeming with fish we no longer have to depend on at all. That is true liberation. This is freedom *from* nature or the logical conclusion of 'decoupling' from nature. We're not there yet, but that's the direction we are traveling. Just look at the rise of plant-based meat alternatives that deliver all the taste of hamburgers or chicken nuggets but without the factory farms. Technological innovation is always the answer.

The early modern prophets of the orthodox faith have taken a lot of grief for basing their doctrines on the parable of the "state of nature." Critics point out that humans have always been technological and social creatures—that there never was a Robinson Crusoe phase of history with the naked individual alone in the wild. That's all true! The fisherman, for example, has a boat and a complex culture. But this misses the point of the parable. The point is to highlight the process of taming the Fates to gain control and security. As Thomas Hobbes wrote in *The Leviathan*, with his own parable about the state of nature, 'security' is "The finall Cause, End, or Designe of men, (who naturally love Liberty, and Dominion over others)" (Part II, Chap. XVII). It is security that the "state of Nature" most sorely lacks. To get it, we must recruit the energies of nature to do our work.

Bibliography

Bacon, Francis. 1620 (1902). *Novum Organum*. New York: P.F. Collier & Son.

Birol, Fatih. 2007. Energy Economics: A Place for Energy Poverty in the Agenda? *The Energy Journal* 28 (3): 1–6.

Böll, Heinrich. 1986. *The Stories of Heinrich Böll*. Leila Vannewitz, trans. New York: Knopf.

BP World Energy Outlook. 2019. *BP Statistical Review of World Energy, 68th Edition.* https://www.bp.com/content/dam/bp/business-sites/en/global/corporate/pdfs/energy-economics/statistical-review/bp-stats-review-2019-full-report.pdf

Chafe, Zoë A., et al. 2014. Household Cooking with Solid Fuels Contributes to Ambient PM2.5 Air Pollution and the Burden of Disease. *Environmental Health Perspectives* 122 (12): 1314–1320.

Epstein, Alex. 2014. *The Moral Case for Fossil Fuels*. New York: Penguin.

Gaye, Amie. 2007. *Access to Energy and Human Development*. UNDP Human Development Report Occasional Paper. http://purocihle.rrojasdatabank.info/gaye_amie.pdf

Hall, Charles. 2011. Synthesis to Special Issue on New Studies in EROI (Energy Return on Investment). *Sustainability* 3 (12): 2496–2499.

Hobbes, Thomas. 1651. *The Leviathan*. Project Gutenberg. https://www.gutenberg.org/files/3207/3207-h/3207-h.htm

Jaczko, Gregory. 2019. I Oversaw the US Nuclear Power Industry. Now I Think It Should Be Banned. *The Washington Post*, May 17. https://www.washingtonpost.com/outlook/i-oversaw-the-us-nuclear-power-industry-now-i-think-it-should-be-banned/2019/05/16/a3b8be52-71db-11e9-9eb4-0828f5389013_story.html

Karlsson, Rasmus. 2015. Three Metaphors for Sustainability in the Anthropocene. *The Anthropocene Review* 3 (1): 21–32.

Keith, David. 2007. A Critical Look at Geoengineering Against Climate Change. *Ted Talk*, September. https://www.ted.com/talks/david_keith_a_critical_look_at_geoengineering_against_climate_change?language=en#t-175956

Lavelle, Marianne. 2013. Five Surprising Facts about Energy Poverty. *National Geographic*, May 30. https://www.nationalgeographic.com/news/energy/2013/05/130529-surprising-facts-about-energy-poverty/

Lomborg, Bjorn. 2001. *The Skeptical Environmentalist: Measuring the Real State of the World*. Cambridge: Cambridge University Press.

———. 2014. Germany's Energy Policy Is Expensive, Harmful, and Short-Sighted. *Financial Times*, March 16. https://www.ft.com/content/9d6ba56a-a633-11e3-8a2a-00144feab7de

Lovelock, James. 2019. *Novacene: The Coming Age of Hyperintelligence*. London: Allen Lane.

Mann, Horace. 1853. *Slavery: Letters and Speeches*. Bedford: Applewood Books.

Matthews, Dylan. 2018. 23 Charts and Maps that Show the World Is Getting Much, Much Better. *Vox*, October 17. https://www.vox.com/2014/11/24/7272929/global-poverty-health-crime-literacy-good-news

Mock, Jillian, and John Schwartz. 2019. What if We All Ate a Bit Less Meat? *New York Times*, August 21. https://www.nytimes.com/2019/08/21/climate/what-if-we-all-ate-a-bit-less-meat.html

Nordhaus, Ted, and Alex Trembath. 2019. Is Climate Change like Diabetes or an Asteroid? *The Breakthrough Institute*, March 4. https://thebreakthrough.org/articles/is-climate-change-like-diabetes

Otto, Mary. 2017. *Teeth: The Story of Beauty, Inequality, and the Struggle for Oral Health in America*. New York: New Press.

Pinker, Steven. 2018. *Enlightenment Now: The Case for Reason, Science, Humanism, and Progress*. New York: Viking.

Prins, Gwyn, et al. 2010. *The Hartwell Paper: A New Direction for Climate Policy after the Crash of 2009*. Institute for Science, Innovation, and Society, University of Oxford, May. https://eprints.lse.ac.uk/27939/1/HartwellPaper_English_version.pdf

Rosling, Hans. 2018. *Factfulness: Ten Reasons We're Wrong about the World – And Why Things Are Better than You Think*. New York: Flatiron Books.

Shellenberger, Michael. 2017. Why I Changed My Mind About Nuclear Power. *TedxBerlin*, November 17. https://www.youtube.com/watch?v=ciStnd9Y2ak

———. 2019. Why Everything They Say About the Amazon, Including that It's the 'Lungs of the World,' Is Wrong. *Forbes*, August 26. https://www.forbes.com/sites/michaelshellenberger/2019/08/26/why-everything-they-say-about-the-amazon-including-that-its-the-lungs-of-the-world-is-wrong/#492533bf5bde

Shellenberger, Michael, and Ted Nordhaus, eds. 2011. *Love Your Monsters: Postenvironmentalism and the Anthropocene*. Oakland, CA: Breakthrough Institute.

Smil, Vaclav. 2017. *Energy and Civilization: A History*. Cambridge, MA: MIT Press.

Watts, Nick, et al. 2019. The 2019 Report of *The Lancet* Countdown on Health and Climate Change: Ensuring the Health of a Child Born Today is not Defined by a Changing Climate. *The Lancet* 394 (10211): 1836–1878.

Yergin, Daniel. 2008. *The Prize: The Epic Quest for Oil, Money, and Power*. New York: Free Press.

———. 2011. *The Quest: Energy, Security, and the Remaking of the Modern World*. New York: Penguin.

8

Factor M

For what is the heart, but a spring; and the nerves, but so many strings; and the joints, but so many wheels, giving motion to the whole body, such as was intended by the Artificer?
Thomas *Hobbes, 1651*

Let's shift now from parable to history. It is time to consult some of the holy texts and saints of the energy orthodoxy. I began with the parable and a brief evolutionary history of *Homo sapiens* to remind us that we are talking about a universal human quest to control Fate. That is crucial to keep in mind: the Anthropocene truly is a shared human story.

However, as I noted, the *formula* needed to make this quest successful does have a particular historical origin. The ideas and the machines that co-evolved with this formula are the products of modern European and North American societies. Though at times inspired by ancient and non-Western traditions, this history mostly occurred in the past four hundred years in the global north. For a good account of this crucial time period, consult Richard Rhodes' *Energy: A Human History* (2018).

© The Author(s) 2021
A. Briggle, *Thinking Through Climate Change*, Palgrave Studies in the Future of Humanity and its Successors, https://doi.org/10.1007/978-3-030-53587-2_8

A formula is a group of mathematical symbols that express a relationship or that are used to solve a problem or make something. The questions that would form the orthodoxy is: How are heat, light, magnetism, motion, and electricity *related* to one another? Just what *are* these: forces, spirits, powers, fluids? Before the orthodoxy, we didn't even have a steady nomenclature, because we didn't understand the underlying infrastructure of reality on which we might hang our language, let alone build our world.

The orthodox formula is an *equation*. It is an understanding of 'energy' as a gold standard that relates seemingly diverse phenomena with one another and permits their ordered conversions back and forth. This is a whole new way of seeing reality. The Hobbes quote above from his *Leviathan* gives just a taste of this imaginative leap to see the heart as a pump, the nerves as springs, and the joints of the body as wheels. Nature is composed of pieces and parts that can be standardized, swapped out, and upgraded. It is a plug and play world.

This mathematization of nature is the skeleton key to gaining control, the thread we can pull on to eventually unravel the Fates entirely. The ultimate version of this formula is Einstein's $E = mc^2$. The central truth of the orthodoxy is its grasp of energy as a universal currency. That something so abstractly bare-bones and theoretical as 'E' can unleash something as big and brute as a mushroom cloud is a testimony to the raw truth of the orthodoxy. Again, we are not talking about a faith in any deities—we are talking about a faith in our own power to understand and control power. This is the secular church of science.

The road to Einstein's precise formula was long and wending. It has had its share of charlatans and hucksters. A favorite of mine is the German-American techno-utopian John Adolphus Etzler. In 1833, he published a breathless tract *The Paradise within Reach of all Men*, which promised the key to harnessing the powers of wind, water, and waves. By 1843, he swore that we would be completely free from labor. There is a "superabundance of power" slumbering in the Earth millions of times greater than "the sinews and nerves" of all humanity could muster. Alas, Etzler met his match in the tropics. But maybe he was more dreamer than swindler. A man ahead of his time. His promise to travel 1000 miles in twenty-four hours seemed fantastical in the 1830s. That's only 42 miles per hour. Yawn.

One way to periodize the history of the orthodox energy formula is to borrow terms from the philosopher of science Thomas Kuhn (1962) and break it up into three phases: the pre-paradigmatic phase (seventeenth and eighteenth centuries), the crystallization of the paradigm (nineteenth century), and then the conduct of normal science within the paradigm (twentieth and twenty-first centuries). As Vaclav Smil (2006) puts it,

> Theoretical energy studies reached a satisfactory (though not a perfect) coherence and clarity before the end of the nineteenth century when, after generations of hesitant progress, the great outburst of Western intellectual and inventive activity laid down the firm foundations of modern science and soon afterwards developed many of its more sophisticated concepts. The ground work for these advances began in the seventeenth century.

The term 'paradigm' is a better way to talk about the 'orthodoxy' (see Frigo 2018). It means a shared way of seeing the world. There is a shared sense of legitimate problems to solve and ways to go about solving them. There is a shared sense of the 'furniture' of the universe. When oxygen displaced phlogiston, for example, this was not just a new theory of combustion. It signaled a whole new paradigm premised on the periodic table of elements and their lawful interactions. Or think of the heliocentric paradigm displacing the geocentric one, or Darwinian evolution displacing creationism.

Recall the point about paradoxes and the elephant—that before counting can begin we must determine what counts *as* something. A paradigm resolves the confusion about what the elephant is. In this case, it tells us that E is a universal currency that operates by certain rules such as the first and second laws of thermodynamics. Once we have that established, we can start solving puzzles using what Kuhn calls "normal science." We can tinker with engines, transformers, generators, and even the metabolic pathways of algae to make them work better. Indeed, sustained and systemic *progress* is only possible once a paradigm is in place.

Our dreamer Etzler with his utopian vision was just about fifteen years too early. He was still writing with the forceful but disorganized ecstasy characteristic of pre-paradigmatic inquiry. His treatise is still helter-skelter and not yet *disciplined*. The darn thing has no methodology section and ends, not with a bibliography, but letters to would-be donors! A method hadn't been established, because we didn't know quite what those 'powers' in nature were. There wasn't yet a proper federal agency with R&D grants to offer. There wasn't even an oil industry with the capital to finance utopian schemes. Energy had not yet been properly discovered. And without E, there can be no social-technical matrix to put it to work.

Soon, though, the pieces would fall into place. The Frenchman Sadi Carnot had laid the groundwork but nobody had noticed. Then suddenly between 1840 and 1880, things began to crystallize in a flurry of insights. Hermann von Helmholtz and Rudolf Clausius in Germany started piecing the picture together at the same time as Lord Kelvin and James Prescott Joule across the channel. The modern energy paradigm—what you insist on calling the 'orthodoxy'!—was born. It just took more patience than Etzler had.

Now, reams of scientific journals methodically report the testimonies of nature. We understand the joints of nature and we have organized knowledge into the *disciplines* that can carve away at those joints.

Ah, but I have done it again and gotten ahead of myself! I want to baptize you, but not throw you in the deep end right away! So, rather than go back to the early formulations of the modern energy paradigm, let's just wade into the shallow waters closer to our own times. Let's start with the new testaments written after the energy paradigm was already in place. We'll dig down to the old testaments later.

*   *   *

By the twentieth century, the concept of energy-as-universal-currency in physics was precisely formulated. Soon, a variety of scientists started to extend the implications of this E from physics to other domains. For example, the chemist Wilhelm Ostwald saw the universal currency as an energetic imperative at work in the evolution of life on Earth. All living organisms obey the same command: "Do not waste any energy, make it useful" (Ostwald 1912). This law spurs evolutionary competitions to appropriate and use ever-higher flows of energy (i.e., otherwise going to waste).

In his 2017 book, *Energy and Civilization: A History*, Smil extends this evolutionary approach to human civilizations, calling them simply "the largest and most complex organisms in the biosphere" (p. 2). He holds that the history of successive civilizations has operated by its own law of natural selection to increase total mass, rate of matter circulation, and total energy flux through the system: "Civilization's advances can be seen as a quest for higher energy use required to produce increased food harvests, to mobilize a greater output and variety of materials, to produce more, and more diverse, goods, to enable higher mobility, and to create access to a virtually unlimited amount of information" (p. 385).

Smil is not the first to read a universal energy law into human history. In his 1954 *Man and Energy*, the physical chemist A.R. Ubbelohde argued that humans have a 'perennial' need to "control energy" or a "need to achieve dominion over matter." He wrote that "A Robinson Crusoe suddenly abandoned now on a desert island would experience much the same sequence of ideas about how to subjugate the world around him as did the original." For him, the urge to control energy is fundamental to human nature. The fisherman's culture is just an earlier stage on the natural, evolutionary path toward the businessman's culture.

A more influential formulation of the orthodoxy was penned in 1943 by the anthropologist Leslie White in his essay titled "Energy and the Evolution of Culture." This essay is a key touchstone for the energy orthodoxy, one that lays out its logic in plain terms. So, it is worth a closer look.

White embraces the philosophical naturalism that William James resisted in his essay on "The Energies of Men." White declares that "natural science" is a redundancy: "All science is natural; if it is not natural, it is not science." Science tells us what is real. If a 'spirit' cannot be rendered into the central energy formula and its laws, then it does not exist. This is a non-negotiable pillar of the energy orthodoxy. "Everything in the universe," White begins, "may be described in terms of energy." What he means by energy is the scientific definition: "the capacity to do work."

This view of energy is true "on all levels of reality." Plants are "biochemical mechanisms" that convert solar into chemical energies. Animals are 'engines' that convert chemical into mechanical energies. These animal energy conversions are what we call behavior. Humans are animals that use energy to behave. Of course, we also have symbol-manipulating behaviors that we call culture, which too is analyzable in terms of "matter-and-motion, i.e., energy."

"The purpose of culture," White says, "is to serve the needs of man." There are two kinds of needs, which correspond to the two factors in any social system: factor E (ends) and factor M (means). Ends in themselves include "singing, dancing, myth-making, forming clubs or associations for the sake of companionship, etc." Means include "the tools, weapons, and other materials with which man provides himself with food, shelter from the elements, [and] protection from his enemies."

White then makes two moves that are central to the orthodoxy. First, he prioritizes the means: "In any social system M is more important than E, because E is dependent upon M." After all, we can only sing and dance if our bodily needs for food, shelter, and defense have been met. First life, then the good life. Again, instrumental rationality (the means) is of utmost importance.

Second, he argues that ends are timeless and universal: "In the development of a culture, moreover, we may regard E as a constant: a men's club is a men's club whether among savage or civilized peoples. Being a constant, we may ignore factor E in our consideration of cultural evolution and deal only with factor M." Again, we are telling a universal human story—all people sing, dance, love, and yearn for freedom from tyranny and toil. The struggle has all been about how to get the means—factor M—to successfully secure those ends. That is the formula modern science hammered out.

People will always desire and need the same set of ends-in-themselves. Yet we can still talk rationally about progress and thereby evaluate different cultures on the basis of how skilled they are at achieving and securing the ends. To determine how far a culture is along the arrow of progress, we only need to look at their factor M, "the material, mechanical means with which man exploits the resources of nature." More advanced cultures are those that have better means for *ensuring* satisfaction of the constant ends. The goal, on White's account, is to make life "more secure, comfortable, pleasant, and permanent."

Recall our parable of the fisherman. In the story, the ends (factor E) are also the same: both fisherman and businessman wish to lounge, play, and dance. The only difference between the two is the amount of energy they command by way of their different technological conditions (factor M). Greater flows of energy have made the life of the businessman, in White's words, more "secure, comfortable, pleasant, and permanent." The root 'urge' motivating progress from the fisherman's culture to the business-man's culture is the universal "human struggle for existence," which "expresses itself in a never-ending attempt to make of culture a more effective instrument with which to provide security of life and survival of the species" (pp. 339–340).

White argues that "The social organization (E excluded) of a people is dependent upon and determined by the mechanical means with which food is secured, shelter provided, and defense maintained." The clause "E excluded" is crucial, because it indicates that new, more efficient, and more productive technologies and increased quanta of energy do *not* create new needs or desires or in any way alter the constant factor E. Improvements in factor M just provide given, constant ends with greater *assurance* and *abundance*.

Thus, we arrive at White's general law of cultural development, which he encapsulates in his own formula:

$$E \times T = P$$

'E' here is not ends but the amount of energy harnessed per capita. 'T' is the technological means of harnessing that energy (later he replaces this variable with 'F' to emphasize that it is the efficiency of the mechanical means that really matters). 'P' is the magnitude of "human-need-serving product" that results or the "total amount of goods or services produced." In other words, P for productivity is another way of saying factor M.

In a later text, White puts things plainly: "culture advances as the amount of energy harnessed per capita per year increases, or as the effi-ciency or economy of the means of controlling energy is increased, or both" (1959, p. 56). "A culture is high or low," he writes, "depending upon the amount of energy harnessed per capita per year" (White 1959, p. 42).

In one word, this is about *more*. Progress means more production, more P. It is building up that insulation that shelters us from the scissors and arrows of the Fates. Whatever paradoxes result from this insulation are first world problems. They are much preferable to a condition where a meager factor M is too weak or thin to protect us from the Fates.

*   *   *

Having established the formula governing progress, White then offers a sweeping history of how this progress unfolded. He divides time into three periods in a way that still dominates histories of energy. For example, Smil (2017) similarly organizes his book into "grand patterns of energy use." Although many overlaps and sub-categories are involved, the basic motif is subsistence to agriculture to industry. That is, the pattern of progress is a set of transitions from human muscles to animate energies to inanimate energies. This approach to history highlights the universality of the story, because whether in India or Canada, ox muscles introduce the same limits on harvests and electricity grids introduce the same kinds of social and material infrastructures.

Stage one is the energy of subsistence found in nomadic hunting and gathering cultures. Smil (2017) notes that human muscles were the only source of mechanical energy for roughly the first 190,000 years of human existence. This long stage one is a world of small scales and handcrafts. Though trade can flourish, long distance travel is rare and mass production is nearly unheard of. Energy flows are severely limited, White notes, because in a "wild food economy, an animal or a plant is of value to man only after it has ceased to be an animal or a plant, i.e., a living organism." Plants and animals are the stores of energy that humans rely on. In a wild food economy, humans also rely wholly on nature to generate, replenish, and sustain those energetic stores, those living batteries.

Stage two happens when humans begin to control these processes via animal husbandry and agriculture. This dates to roughly 12,000 years ago. The use of animal muscles as extra-somatic forms of energy quadrupled the peak prime mover capacity under our command. Yet the control of plants is the much larger energy pay-off, because grazing animals still depend on the wild food economy of natural ecosystems. As such the "pastoral man" White writes is "still dependent on the 'forces and caprices of nature,'" that is, the Fates.

Of course, stage two can be broken down into numerous sub-stages. Marcel Mazoyer and Lawrence Roudart (2006), for example, frame agriculture as a form of mutualism between an exploiting species (humans) and exploited species (crops). They tell a story of accelerating revolutions as new managed ag-ecosystems emerge, each one expanding the productivity of the land. By the year 2000, the speed of change was incredible: an average acre of cropland received about ninety times more energy subsidies than it did in 1900. Suddenly, potatoes were made of oil. Check that, our world was made of oil.

Stage two is a convoluted process of developing better farming implements, irrigation, and especially fertilizers. Progress is measured in terms of P-for-productivity: yields per unit time or unit of labor or area of land (i.e., the efficiency and power of factor M). A small waterwheel manned by 10 people in Renaissance Europe, for example, could grind enough grain in a day to feed 3500 people—a feat that would otherwise take 250 laborers with hand milling. Smil (2017) calculated the labor productivity for American wheat farmers in terms of minutes required to produce a kilogram of grain. In 1800, it was 7.2 minutes. By 1900, it was a little less than thirty seconds. By 2000, it was six seconds.

How low can this number go? The limiting factor is not land per se, because although that is scarce it can be used over and over again without diminishing its amount. Rather, the key is fecundity or low entropy energy. Ultimately, then, value stems not from land but low entropy energy. Thus, the question is how long the second law of thermodynamics, that is, the decline of available or useful energy, can be forestalled (see Georgescu-Roegen 1986).

Growth and accumulation of wealth began to ramp up during the second energy stage, but there were still strict limits imposed by the primary producers of photosynthesis, animal muscle, human labor, waterwheels, and wind power. It was a wood-powered world with rapidly disappearing forests. In 1700, Massachusetts was 85% forested. In 1870, that figure was 30%. The Royal Society's first published book, *Sylva* (1664), analyzed the scarcity of large trees and implications for the Royal

Navy. This was also a world powered in large measure by the moral abomination of slave labor. As many as 55 million slaves were transported from Africa to America in the Atlantic slave trade: "Every lump of sugar dropped in British teacups depended on shackled muscle" (Nikiforuk 2012, p. 13).

The next energy transition gave birth to stage three, our present era. It is the age of "inanimate energies," especially fossil fuels and nuclear power. Now acceleration truly kicks in. It took about 8000 years to get an order of magnitude increase in primary productivity from animal muscle energy to the output of a good horizontal waterwheel. It took another thousand years or so to get another order of magnitude gain with the steam engine. It took only about 200 more years to get to nuclear power, which adds not just one more order of magnitude but a million-fold increase in power. Human population and standards of living soared as a result of this flood of available energy.

The scarcity of wood for smelting iron prompted the large-scale use of coke (a solid fuel made by heating coal). More generally, the limits of phytomass fuels led to the age of inanimate prime movers. The coal-fired heat engine catalyzed a chain reaction as coal-powered engines drained the mines that made yet more coal available. Now oil and gas enable the technologies that make unconventional shale reserves of oil and gas available. In this way, proven reserves keep growing even as we consume ever-more fossil fuels. The winning formula is a recipe for energy begetting energy.

With the transition to stage three and less dependence on wood, reforestation could occur. Further, as inanimate productive powers increased, the human slave economy became obsolete. Fossil-fueled machines made human slaves simply unprofitable. In his 1849 essay, "Civil Disobedience," Thoreau chastises those who worry about slavery but *do* nothing to stop it: "There are nine hundred and ninety-nine patrons of virtue to one virtuous man." Yet slavery is solved through better technologies far more effectively than it could be through protest or better morals. So too, civil disobedience against climate change may be part of the solution, but it is dwarfed in importance by technological fixes. The young climate strikers would do well to earn engineering degrees.

This three-stage history is instructive but we need to flag what it tends to obscure. First, of course, there are tremendous overlaps and complexities and only about one-fifth of the present human population leads lives fully immersed in stage three. Second, the transitions between stages are rapidly accelerating. Stage one was 190,000 years. Stage two was 12,000 years. Stage three has been only about 200 years and already we face climate change and the challenge of the biggest energy transition to date, the transition to stage four—a decarbonized global economy.

Third, much was said above about energy and forgetfulness, but we left out the biggest point: we forget the hardships of stages one and two! In stage one, food is basically the only energizer. In the Roman Empire, an advanced stage two society, food accounted for 45% of total energy. Even in the globe's most developed economies in the 1820s, food was still about 30% of all energy. Yet by 1960, food was less than 2% of the total energy supply for human civilization.

In 1776 when Thomas Jefferson penned the Declaration of Independence, the world's prime mover capacity was basically the same as it had been for thousands of years. Smil (2017) estimates that 97% of total power was provided by people and animals. In 1850, the animate labor of bodies still provided about 80% of total power. By 1910, that number had dropped to 60%. Now it is roughly 1%. All the rest of the world's available power is installed in internal combustion engines and electricity generators—machines that didn't exist a handful of generations ago. It's the Great Acceleration.

Finally, this three-stage history obviously doesn't tell us about the present and future fourth stage. In 2006, the US Department of Energy commissioned a study about solar energy (Tsao et al. 2006). It began by noting that global energy consumption rates will *triple* by 2100 from about 13 terawatts (TW) to 43 TW. If we stay in stage three with a factor M of mostly fossil-fueled machines, this will *quintuple* the atmospheric concentration of CO_2. Thus, the need for another revolution, another wave of progress to what the report calls "C-neutral power."

The report then turns to the theoretical potential of solar energy and calculates that there is "more energy striking the earth's surface in one and a half hours…than worldwide energy consumption in the year 2001 from all sources combined" (p. 10). The energy orthodoxy supplies the formula for turning theoretical potential into reality. This can be seen in the plummeting prices of solar panels and battery storage technologies. This is driven by new scientific discoveries about how to boost solar cell efficiency as well as technological breakthroughs such as cadmium telluride solar panels. Some analysts support the Solar Dominance Hypothesis, which argues that solar power plus battery storage technologies are "fundamentally disruptive, not incremental technologies" (Goodstein and Lovins 2019). Solar power may rapidly transform the energy economy in the coming two decades.

Stage four is happening in even more radical ways, too. Solar panels are just technological leaves doing a different kind of photosynthetic energy conversion. The original leaves, the greenery of the Earth, represent an even more profound opportunity for tapping into global solar influxes. Indeed, leaves are *the* prime mover of the Earth's energy economy. In their natural state, however, this factor M suffers from severe efficiency problems. Most plants only convert solar energy into useful chemical energy with an efficiency of 1%.

Traditional plant breeding seems to have hit a "yield ceiling" and is unable to boost efficiency. But targeted genetic manipulations in photosynthetic and metabolic pathways show promise that we can boost the productivity of plants significantly. Some trials have increased energy conversion efficiencies to over 20% (see Orr et al. 2017). In this way, gene-edited plants can be coaxed into powering the growing human economy in a way that rebalances the carbon cycle. Such direct control over plants represents a radical boost to factor M. Indeed, it is one of the most powerful ways we can loosen the grip of the Fates.

Bibliography

Etzler, John Adolphus. 1833. *The Paradise within Reach of all Men, without Labor, by Powers of Nature and Machinery.* London: John Brooks.

Frigo, Giovanni. 2018. *An Ecocentric Philosophy of Energy in a Time of Transition.* Unpublished dissertation, University of North Texas.

Georgescu-Roegen, Nicholas. 1986. The Entropy Law and the Economic Process in Retrospect. *Eastern Economic Journal* 12 (1): 3–25.

Goodstein, Eban, and Hunter Lovins. 2019. A Pathway to Rapid Global Solar Energy Deployment? Exploring the Solar Dominance Hypothesis. *Energy Research & Social Science* 56. https://doi.org/101197.10.1016/j.erss.2019.05.007.

Hobbes, Thomas. 1651. *The Leviathan.* Project Gutenberg. https://www.gutenberg.org/files/3207/3207-h/3207-h.htm

Kuhn, Thomas. 1962. *The Structure of Scientific Revolutions.* Chicago: University of Chicago Press.

Mazoyer, Marcel, and Laurence Roudart. 2006. *A History of World Agriculture: From the Neolithic Age to the Current Crisis.* New York: New York University Press.

Nikiforuk, Andrew. 2012. *The Energy of Slaves: Oil and the New Servitude.* Vancouver: Greystone Books.

Orr, Douglas, et al. 2017. Engineering Photosynthesis: Progress and Perspectives. *F1000Research*: 1891. https://doi.org/10.12688/f1000research.12181.1.

Ostwald, Wilhelm. 1912. *Der energetische Imperativ.* Leipzig: Akademische Verlagsgesselschaft.

Rhodes, Richard. 2018. *Energy: A Human History.* New York: Simon and Schuster.

Smil, Vaclav. 2006. *Energy: A Beginner's Guide.* Oxford: Oneworld Publications.

———. 2017. *Energy and Civilization: A History.* Cambridge, MA: MIT Press.

Thoreau, Henry David. 1849 [1903]. *On the Duty of Civil Disobedience.* London: The Simple Life Press.

Tsao, Jeff, et al. eds. 2006. Solar FAQs. https://www.sandia.gov/~jytsao/Solar%20FAQs.pdf

Ubbelohlde, A.R. 1954. *Man and Energy.* Baltimore: Penguin Books.

White, Leslie. 1943. Energy and the Evolution of Culture. *American Anthropologist* 45 (3): 335–356.

———. 1959. *The Evolution of Culture: The Development of Civilization to the Fall of Rome.* New York: McGraw-Hill.

9

Putting Descartes Before the Horse

> *[A]nd thus make ourselves the masters and possessors of nature.*
> Rene Descartes (*1637*)

You are ready now for a deeper dive into the modern energy paradigm. We know it is about controlling fate. And we know this is a perennial human quest that has been accelerated by a formula derived in Europe and North America across the last four hundred years. Finally, we know that progress in this quest is measured by a more powerful and efficient factor M. Of course, that 'M' is for means but it also signifies 'more.' More energy. More ability. More wealth. More choice. More happiness. Above all, more *security*.

What we don't know yet is how this winning formula was derived. Before the modern paradigm of Einstein's E there was a pre-modern paradigm, or perhaps simply a pre-paradigm. Most energy histories are about the machines—the steam engine, the railroad, the automobile, the dam, the electric motor, the nuclear reactor, and so on. Yet as Lewis Mumford (1934) noted, behind all those material inventions, there "was not merely

© The Author(s) 2021
A. Briggle, *Thinking Through Climate Change*, Palgrave Studies in the Future of
Humanity and its Successors, https://doi.org/10.1007/978-3-030-53587-2_9

a long internal development of technics: there was also a change of mind" (p. 3).

To put the cart before the horse is to bungle the proper order of things. But to put Descartes before the horse—or rather, the horseless carriage—is to set things straight. It was Descartes' new state of mind that necessarily preceded industrialization. I use René Descartes as an exemplar of a new way of thinking—he is a model of the modern mind. Putting Descartes before the horse means getting at the revaluation of values that cleared the way for the formula of E and the modern world. Without this shift in worldviews we don't get that revolutionary third stage of energy history discussed in the last chapter and, of course, there would be no hope for the genius required to transition to a decarbonized global economy.

So, there was a change of mind. A change *from* what *to* what, though? There is no single answer to that question. For example, Galileo Galilei—another exemplar of the modern mind—initiated a change from Earth as unique and central to Earth as another cosmic body. There was also a reversal in social ontology, from holism to individualism (Dumont 1986). A high-energy world is full of individuals concerned about the stifling claims of place and society and convinced of their *own potential energies!* By breaking the constricting ties of place, season, and tradition, high-energy machines help to grant individuals their proper due, that is, to prioritize the individual over the whole.

Above all, we are talking about a change from organism to mechanism. It is a shift to a clockwork universe. The pre-modern view of energy, *kinesis*, was of a process of growth and life unfolding according to rhythms that were unique to each *kind of thing*. Descartes, by contrast, helped inaugurate *a world revealed as* objects that can be quantified and controlled. The new *doxa* is energy like "the movements of a clock" where the body-machine works according to "the arrangement of its counterweights and wheels." With Descartes, all bodies are *res extensa* splayed across abstract coordinates of space-time. This view of bodies allows for cutting and pasting, moving and re-ordering. It allows for a shift from the "speculative philosophy" of the ancients to what Descartes called a "practical philosophy" where experiments produce results that others can build on to progressively master fate.

The clock is the key. It was at least as important as coal to the age of industry. By the fourteenth century, the Tyne Valley in northern England was largely owned by abbeys and monasteries. Coal was so abundant that it crumbled like pie crust from the river banks. There was, however, something of a theological dilemma. There was a paradox—what *is* coal? Shall we see it *as* Mother Earth or *as* resource?

In keeping with long-held traditions, the monks often conceived of the Earth as a feminine partner of the divine creator (Merchant 1989). Pliny the Elder's *Natural History* (ca. 77 CE) portrayed mining as an invasive assault on the body of Earth. A generation earlier, in his *Metamorphoses*, Ovid bemoaned the way miners "dug into her vitals." These sentiments were carried forward by theologians who commonly considered mining immoral, a sin "not only against Mother Earth but also against divine will" (Northcutt 2013, p. 36). If God gifted coal in accessible chunks, that is one thing. To hack it out of the ground ourselves, however, is another—it is to rebel against the divine order where everything has its proper place.

The theological winds, though, were shifting. William of Ockham (1287–1347) disassociated the will of God from the 'accidents' of the Earth, arguing that there is only an arbitrary relationship between the physical order and the divine moral order. Earthly existence is contingent, prone to swerving this way or that way (Greenblatt 2011). If that's the case, there can be no theological objections to rifling through the bowels of the Earth and rearranging the pieces. The 'univocity' of Duns Scotus (1266–1308) also played a role. He maintained that certain predicates could be applied with the same meaning to God and creatures. Humans, after all, are made in the image and likeness of God, so we must share something essential. The differences between humans and God are in *degree not kind*. As co-creators, it is our prerogative to change nature.

Perhaps more important than any theological doctrine, however, were the clocks on the walls of the medieval monasteries where such doctrines were discussed. Mumford (1934) argued that the monks unwittingly crafted a new cosmology with their clocks. Clouds could baffle the sundial and night frost could stop the water clocks. But the tick-tock of abstracted time would continue the same regardless of place or conditions. With the clock, standardization was born. Regular units of time stood out in their abstracted sameness and could be counted by unchanging metrics.

This helped to ease the birth of the modern scientific worldview, which is also a view of lawful regularities and hidden equivalencies. For example, F = ma holds true for the cannonball, the sparrow, and the entire planet. In all cases, the equation about *force* is the same just as the tick-tock of the clock is the same whether at night or day, whether in France or China or on the space station. This is the worldview that made the discovery of E (a ruling equivalence under all apparent difference) possible.

The hand of the clock pierced the pre-modern organic world of form and proportionality. Mumford depicts clock time as the mother of modern machines:

> When one thinks of the day as an abstract span of time, one does not go to bed with the chickens on a winter's night: one invents wicks, chimneys, lamps, gaslights, electric lamps, so as to use all the hours belonging to the day…Abstract time became the new medium of existence. Organic functions themselves were regulated by it: one ate, not upon feeling hungry, but when prompted by the clock: one slept, not when one was tired, but when the clock sanctioned it.

The clock became the first machine of modernity. The new mechanistic philosophy that Descartes helped to birth "explained phenomena in purely physical or deterministic terms." Modern machines are fueled by a mechanistic view of energy as E. Before they could be revved up, the premodern worldview had to be destroyed, which meant an assault on Aristotle and his *energeia*. Indeed, it meant cracking open the "closed world" to fashion the possibilities of an "infinite universe" (Koyré 1957).

* * *

Some 2500 years ago, the first philosophers wondered about change. Life and death. Growth and decay. Rising and falling. Freezing and thawing. They thought that despite all this change there must be something permanent, something "which is primary, which persists, which takes various forms and undergoes this process of change" (Coppleston 1993, p. 20). There is some *Urstoff* that cannot be created or destroyed, something that undergoes the change but is not changed. The first philosophical schools were split by their answers to the question of just what this 'stuff' is. Water, perhaps? Or fire?

Parmenides adopted a different approach. The core of it is as simple as it is counter-intuitive: Becoming, or change, is an illusion. If something comes to be, it either comes from being or not-being. If it comes from being, then it already is, so it did not *become*. And it cannot come from not-being, because out of nothing comes nothing. Being simply *is*. This is what reason dictates, and only reason (not the senses) can apprehend Reality.

Parmenides' disciple Zeno created riddles to prove the impossibility of motion. Most famously, he imagined a race between Achilles and the tortoise. Being a good sport, Achilles gives the tortoise a head start. When Achilles reaches the place where the tortoise started, the tortoise will be further along. When Achilles reaches *that* place, the tortoise is further along still, and so on. If space is composed of lines with an infinite number of points, then Achilles would have to traverse an infinite distance and will never catch the tortoise. Therefore, motion is impossible.

But that can't be right. Born about a hundred years after Zeno, Aristotle sought to give a rational account of the becoming (change) that appears to our senses. He did so with potentiality or *dynamis*, a kind of halfway house of Being. Something can be X actually but Y potentially: "It *is* X, but is going to be Y in the future in virtue of a potentiality, which is not simply nothing, yet is not actual being" (Coppleston 1993, p. 52). The acorn is the not-yet-oak-tree or at least has this not-yet power inside of it.

All of this, so far, is well and good. Indeed, these pre-modern philosophers rightfully imagined a law-governed universe to replace the kooky antics of Zeus and his capricious mythological minions. In this way, they started down the path toward the scientific cosmology of modern physics and the energy paradigm.

But they took a wrong turn: the natural laws they imagined applied uniquely to each kind of thing. Recall the theological dilemma about coal and the *proper place* of things in the natural world. This was a hangover from the pre-modern paradigm where each thing has its own *nature* that determines its rightful limits and proper fit. They had not yet learned that nature is a mechanism composed of standardized parts and governed by lawful tick-tock regularities underneath all apparent differences (that DNA, for example, was the code for each kind of thing and that it could be spliced across species). Aristotle believed in natural kinds. It would take not just Descartes, but Darwin to demonstrate that there is no such thing. Monkeys become humans. All is flux. No lines demarcating proper from improper can be drawn on the basis of *nature*.

Aristotle's concept of *energeia* was at the heart of this confusion. In order for a potentiality to be manifest, *energeia* is needed. *Energeia* as a noun meaning 'activity' or "the state of being busy or at work" is Aristotle's coinage, something he developed from the adjective *energos* for active or at work. For Aristotle, the essence of a thing is a kind of perpetual work (see Sachs 2018). The fish resting at the bottom of the ocean is *actively making* itself as it metabolizes. Even the rock next to the fish is straining toward the center of the Earth as is in keeping with its nature. All things are busy *thinging* in their own ways through continuous expenditures of effort.

To be something is always to be at work *in a certain way*. *Energeia* is about the activities that are fitting or appropriate to any given kind of thing. Each thing has its own *energeia* that cannot be swapped out for some other kind. This allows the concept to be simultaneously descriptive and prescriptive; how the thing is *and* how it should be. Crucially, this ontology (this *doxa* or view of reality) puts *limit* at the center of things. There are only certain kinds of activities appropriate to, say, the fish or the rock.

Given that it is about norms and limits, Aristotle's world of *energeia* is full of judgments about 'goodness' understood as what fits or what is proportionate or appropriate as well as excellent. Is the fish *fishing* (being-at-work-as-fish) in a good way? Is that an excellent example of a watch *watching* or a knife *knifing* or a kite *kiting* or a cat *catting*, and so on? (The word related to *energeia* that captures this is *entelecheia* or being-at-work-staying-itself.) We can always imagine the most excellent version of any given kind of thing and compare each particular thing-at-work with that ideal model where its unique potential energies are realized to utmost perfection.

At the core of this pre-modern energy paradigm are the notions of form and function. Each thing's peculiar *energeia* is fitted to its form and directed toward fulfilling its *telos*—its function or purpose. The good watch keeps accurate time. The good knife cuts cleanly. The good football player is swift, strong, and makes smart decisions under pressure. These are the *virtues* of any given thing.

What about the good human being—what is the proper form and function of our *energeia*? In the *Nicomachean Ethics*, Aristotle says that it cannot simply be living ("the life of nutrition and growth"), because that we share with plants. And it cannot be locomotion and sense perception, because we have that in common with other animals. The human perfection, then, must be "an active life of the rational element" (1098a). Aristotle argues that: "the good of humans is an activity (*energeia*) of the soul in conformity with excellence or virtue (*arête*)" (1098b 15). That activity is *eudaimonia*, which means happiness or flourishing. Happiness is the *energeia* of a human being a virtuous human.

The virtues, Aristotle argues, are formed by habit or *ethos*. This means they are not "implanted in us by nature, for nothing which exists by nature can be changed by habit" (1103a 15). You can't train a stone, which has a natural downward movement, to become habituated to moving upward. But you can train humans to habitually behave virtuously, so "we are equipped by nature with the ability to receive" the virtues. They are in us as a *dynamis* (potential). We realize them (make them real) by putting them into action through discipline and practice. As Plato noted, this kind of energy is focused on wrestling with our soul horses more than wrestling with the forces of the external natural world.

This view of 'energy' as virtue is fine on a personal level. The trouble starts when this energy paradigm becomes a political project, that is, when the state or society attempts to shape individuals into the 'proper' form with the 'right' virtues. This was the 'perfectionism' of pre-modern politics. Happiness, Aristotle argued, is *not* just whatever we happen to take pleasure in: "happiness cannot exist apart from virtue." There are different kinds of activities or *energeia*. Some are higher or more fitting for the kinds of creatures we are and some are lower and inappropriate. Since there are morally better and worse activities, the same holds for

124   **A. Briggle**

pleasures: "The pleasure proper to a morally good activity is good" (1175b 25).

Since the virtuous or excellent person is "the measure of each thing," then what seems to him or her to be pleasant is in fact pleasant. Of course, other things will seem pleasant to those who are "corrupted and perverted" but "such things are not actually pleasant" (1176 15-20). In other words, we can be wrong about what we should prefer and who we are *supposed to become* and the job of the state is to correct us and set us on a higher path.

The question is: are you morally good and excellent or are you "corrupted and perverted"? Do you like car racing, football, bitcoin mining, international travel, and so on? Should your pleasures count as examples of proper *energeia*? Who is to decide and by what standard? Aristotle begs the question by asserting that the virtuous man is the measure of each thing. Who called him virtuous in the first place!? Couldn't someone else claim that their activities and pleasures are the true virtues and, thus, the appropriate standards for human behavior? When that happens, how could we ever decide who was right?

In this way, Aristotle's so-called natural science can be unmasked as a cover for his own preferred conventions. It is a form of elitism and ableism; a mold that some can be crammed into, but many others will never fit, including women and "natural slaves." Worse, the pre-modern energy paradigm is a recipe for religious wars about who has the right virtues—who has the formula for realizing the human *dynamis* in the proper way. Descartes saw that the way out of this was to let go of the last vestiges of myth, that notion of a proper form or function (*telos*) to humanity. Or as Hobbes said, there is no *summum Bonum*, no greatest good or guiding ideal of excellence. This insight would eventually wend its way to a more modern formulation of freedom: Let each person do what they want to do, so long as it does not harm others.

This plea for toleration became the basis of the modern political project. Animating it was a new sense of energy—not the activity of fulfilling a proper essence or fitting into one's proper place, but a neutral enabler to help you become whoever you want to be!

*   *   *

On June 21, 1776, Thomas Jefferson sent a draft of the Declaration of Independence to Benjamin Franklin. Dr. Franklin made one crucial edit. Jefferson had written: "We hold these truths to be sacred and undeniable." It was Franklin who crossed out the last phrase and substituted "self-evident." With a stroke of a pen, Franklin changed the founding American principle of natural rights from an assertion of religion to an assertion of rationality. He clarified the energy paradigm behind life, liberty, and happiness. We are not talking about pursuing some proper essence of happiness, one which the king or priests would police. No, we are talking about individuals freely pursuing *their own chosen ideal of happiness*.

By then, Franklin had been a member of the Royal Society for over twenty years. He had been awarded the Copley Medal, the Society's highest honor, for his famous kite experiment in a thunderstorm. Technologically, Franklin's experiments were a major step in the quest to harness the forces of nature. They helped to move electricity from the status of parlor tricks to serious applications.

The technology, though, wasn't the real source of power. That comes from the idea behind the machines. Franklin's "discoveries in electricity" were so important, because they uncovered a hidden equivalence in nature. They showed that the little shocks delivered by Leyden jars or "electrified bumpers" (full glasses with static electricity to tingle the lips) were identical with the sky-shattering power of lightning bolts. Joseph Priestley raved about the experiment, because it demonstrated "the perfect similarity between electricity and lightning" (in Rhodes 2018, p. 170). This "perfect similarity" was E: a new kind of energy crossing all boundaries between supposedly natural kinds. It is a formless, aimless shape-shifter. It is *not* Aristotle's *energeia* with its limits imposed by the form and function of various natural kinds. No, E is the universal currency of a cut and paste world, the skeleton key to unlocking productivity.

Inspired by Franklin, the Italian surgeon and physiologist Luigi Galvani wondered if this "perfect similarity" might also be at work in the bodies of animals. Could the stuff of lightning and Leyden jars also be at work in the grease-coated nerves that traverse the muscle tissues? In the 1790s, shortly after Franklin's death, Galvani and his wife Lucia began their electrical experiments with frog legs. They attached the crural nerve to a Dollond machine, which produced "artificial electricity." They even used Franklin's "natural electricity" from stormy skies to shock the dissected frog legs into quivers that eerily resembled life.

The image of lightning re-animating dead tissue sparked Mary Shelley's imagination. 'Galvanism' makes one crucial appearance in her 1818 *Frankenstein*. Dr. Frankenstein's ghastly quest begins when he hears a lecture on electricity that changed his life. "All that had so long engaged my attention grew suddenly despicable," Frankenstein says, "I at once gave up my former occupations, set down natural history and all its progeny as a deformed and abortive creation, and entertained the greatest disdain for a would-be science which could never even step within the threshold of real knowledge."

Shelley's prose conjures the words of Descartes who wrote in his 1637 *Discourse on Method*, "As to the other sciences, inasmuch as these borrow their principles from philosophy, I judged that no solid superstructures could be reared on foundations so infirm." The old foundations of knowledge—whether from Aristotle or the Church—had to be razed. A new philosophy was aborning. This scientific revolution was behind Franklin's portentous edit to the American Revolution where the evidence of common sense replaces the mysterious authority of the sacred. The "perfect symmetry" of energy captures the essential point: nature is governed by laws that operate the same whether on Earth or in space, whether in the skies or in the nerves, whether in frogs or humans. Everything is interchangeable and energy was the universal currency of conversion.

Galvani, however, still had one foot in the world of Aristotle. He believed that animals generate a special kind of electric charge in their muscles and transmit it through their nerves: "this 'animal electricity' was believed to be different from static electricity or lightning and *unique to each animal*. Galvani believed his experiments had elicited animal electricity rather than ordinary, non-vital electricity" (Rhodes 2018, p. 174, emphasis added). Galvani was still looking at things through the old paradigm of natural kinds, each with a unique *energeia* fitted to it.

Along comes Allesandro Volta, a professor of physics at the University of Pavia and member of the Royal Society. Volta doubted the whole theory of a unique animal electricity. After all, why did the leg only respond when linked to a bimetallic conductor? He believed the frog's response came from the contact between the two dissimilar metals, not from the frog itself. He showed how the electric current did not originate in the muscle. He even got the leg to spasm by clipping on an arc of tin leaf and brass without the Leyden jar. If he could get this reaction from a bimetallic conductor, Volta concluded, "then there is surely no reason to assume that a natural, organic electricity is at work here" (in Rhodes 2018, p. 175).

(Note, though, that even when animated by this universal currency from an 'unnatural' source, the frog leg still kicks as a frog leg. It still does its frogly thing rather than, say, flying like a wing or quivering like a leaf. Perhaps Aristotle cannot be entirely dismissed (Brown 2019).)

Galvani did not swear Napoleon's loyalty oath in 1796 (he was a devout Christian and this would have committed him to atheism). He died in poverty. Volta did take the oath and in 1800 he introduced in a letter to the Royal Society his *pila*, a pile of dissimilar metals, copper or silver, tin or zinc, stacked together separated by saltwater-saturated pieces of cardboard. He had recreated the saline, electric body of the frog without the frog. The modern chemical battery was born as a frogless frog jumping with E.

Hans Christian Ørsted, a Danish physicist, would soon discover electromagnetism and further refine the notion of a "perfect similarity" in the concept of a grand "unity of nature." Electricity, magnetism, heat, light—all fundamentally the same. The clockworks of nature were being discovered, understood, and manipulated. Manipulated for what purpose? The notion of *telos* may have died with Aristotle's *energeia*, but a new purpose arose along with E to take its place: *convenience*.

Bibliography

Brown, Keith Maggie. 2019. "Queering and Wyrding." *Call Me Maggie*. https://callmemaggie.com/2019/06/13/queering-and-wyrding/

Copleston, Frederick. 1993. *A History of Philosophy, Vol. 1, Greece and Rome from the Presocratics to Plotinus*. New York: Doubleday.

Descartes, Rene. 1637. *Discourse on the Method on Rightly Conducting One's Reason and Seeking Truth in the Sciences*. Project Gutenberg. https://www.gutenberg.org/files/59/59-h/59-h.htm

Dumont, Louis. 1986. *Essays on Individualism: Modern Ideology in Anthropological Perspective*. Chicago: University of Chicago Press.

Greenblatt, Stephen. 2011. *The Swerve: How the World became Modern*. New York: W.W. Norton.

Koyré, Alexandre. 1957. *From the Closed World to the Infinite Universe*. Baltimore: Johns Hopkins University Press.

Merchant, Carolyn. 1989. *Ecological Revolutions: Nature, Gender, and Science in New England*. Chapel Hill: University of North Carolina Press.

Mumford, Lewis. 1934. *Technics and Civilization*. New York: Harcourt.

Northcutt, Michael. 2013. *A Political Theology of Climate Change*. Grand Rapids, MI: Wm. B. Eerdmans Publishing Co.

Rhodes, Richard. 2018. *Energy: A Human History*. New York: Simon and Schuster.

Sachs, Joe. 2018. Aristotle: Motion and Its Place in Nature. *Internet Encyclopaedia of Philosophy*. http://www.iep.utm.edu/aris-mot/#H2

Shelley, Mary. 1818 (1994). *Frankenstein*. London: Dover.

10

Convenience

*Civil Engineering is the art of directing the great Sources of Power in Nature
for the use and convenience of man.*
Thomas Tredgold (*1828*)

When our minivan is in need of gas, we head for the local convenience store. There, all manners of commodities have been brought together. Not just the gasoline, but also beef jerky, chips, soda, windshield washer fluid, magazines, bobble-head dolls, and much more. Imagine running time in reverse and watching, say, that bag of potato chips jump back in the box, back on the truck, and back to the factory. Imagine watching supply chains working in reverse like fishing lines pulling the potatoes, the tinfoil bags, and the ink printed on the bags back to their origins until eventually all those pieces pull apart and disappear into far-flung pockets of the Earth.

If you think about it, it is remarkable that all of that *convenes* in the store. Of course, you don't have to think about it if you don't want to! Thanks to machines that harness E, it just happens. Presto.

© The Author(s) 2021
A. Briggle, *Thinking Through Climate Change*, Palgrave Studies in the Future of
Humanity and its Successors, https://doi.org/10.1007/978-3-030-53587-2_10

Now go back further in time to Scotland in the mid-eighteenth century when average life expectancy is south of forty years. It is a hard life of farming, herding, cold, and scarcity. There is no convenience store with bags of chips or cans of propane. There is not enough energy to break the bonds of place to pull the strands of the globe into a network of protection against fate.

It was in that world where David Hume was pondering energy. He had embraced the Newtonian mechanistic worldview and rejected the old explanations of "*power, force, energy* or *necessary connexion.*" We know, he wrote in his 1748 *An Enquiry Concerning Human Understanding* that the motion of our body follows upon the command of our will. But we do not understand "the energy, by which the will performs so extraordinary an operation." The mechanical view introduced a dualism between mind and body, which created a perplexity: How does the mind or spirit or E move a material body?

From the mechanical point of view, the particular body under investigation does not matter. Whether it is human limbs, billiard balls, or meteorites, we can see the conjunction of events. The cue ball moves, then the eight ball moves. The lightning strikes, then the frog leg twitches. But the "energy of the cause," Hume said with exasperation, is unintelligible. For human actions, he figured that reason cannot be the prime mover. Rather, "Reason is the slave of the passions." Reason does not motivate us or determine ends, only desire does that. Passions cannot be either reasonable or unreasonable, because they are "original existences" like E. We cannot be criticized for our desires: it is "not contrary to reason to prefer the destruction of the whole world to the scratching of my finger." Aristotle's morality of *energeia* attuned to "correct desires" is long gone.

What can take its place? Hume saw the answer. He said that the morality of a mechanical world is one based on utility and convenience. If energy is not seeking some 'natural' or proper fit, then it needs to be guided by our preferences. His essay on "The Stoic, Or the Man of Action and Virtue" begins:

> nature has endowed man with a spirit and placed him in a situation that urges him, by necessity, to employ, on every emergence, his utmost art and

industry.... where nature furnishes the materials, they are still rude and unfinished, till industry, ever active and intelligent, refines them from their brute state, and fits them for human use and convenience. (*Essays, Moral, Political, and Literary*, Pt. 1, Chap. 16, 1777)

Hume's views would influence the young Thomas Tredgold to offer the above definition of engineering at the start of this chapter, which was adopted by King George IV in the Royal Charter given to the Institution of Civil Engineers (Mitcham 2019). In this way, 'convenience' was given what no religion could have in a modern society: the official seal of approval as the national way of life. The engineers were chartered to direct the energies of nature ("the great Sources of Power") in the making of a more convenient world. They would put the pieces together to suit our desires.

The Latin *convenientia* means "meeting together, agreement, accord, harmony, conformity, suitableness, [or] fitness." In the pre-modern paradigm, something could be described as convenient if it was in accord with nature, or a morally appropriate fit. Indeed, *energeia* was all about this kind of fittingness or proportionality. The Aristotelean virtues are about striking the right balance, finding the mean. But the mechanical worldview lent a radically new meaning to convenience as ease or the absence of trouble. Thomas Tierney (1993) argues that the different meaning has to do with how 'suitable' (fit or proper) has shifted and narrowed:

Convenience is no longer a matter of the suitability of something to the facts, nature, or a moral code; suitability in the modern meaning of convenience refers back to the person, the self. Something is a convenience or convenient in the modern sense of these words if it is suitable for personal comfort or ease.

The new meaning of 'convenient' signals the revaluation of values that attends to a paradigm shift. The pre-modern world of rightly proportioned bodies gives way to the modern world of bodies seeking to overcome all limits and burdens. Gone is Heidegger's "gathering" of the fourfold (Earth, sky, mortals, and divinities). We are not convening with our nature. We are convening around us a world of ease and availability.

There is not a fourfold or one right way of convening but millions of ways—as many ways as people find, well, convenient! There is no need for a limiting doctrine of the mean, because more experiences and commodities can always be made more comfortably available.

Now we are uncovering more layers of the fisherman or "state of nature" parable. Hobbes describes the state of nature as "incommodious" and offers the social contract as a pathway to "commodious living" (*Leviathan* [OUP 1909, p. 98]). Yet the best formulation for convenience comes from John Locke, the man who did the most to inspire Jefferson's revolutionary pen. Locke says that "civil government is the proper remedy for the inconveniences of the state of nature" (Second Treatise, ch. 2, sec. 13).

Of course, Locke doesn't figure prominently in traditional histories of energy that focus on machines and their inventors. But, like Descartes, Locke cleared the conceptual ground for the building of the modern world. He starts with an analysis of "the labor of our hands." Labor is in some sense the lowliest form of our energies. It is based on the appropriating activity of metabolism. Yet Locke finds in labor our highest potential and the key to the good life.

*   *   *

In his 1609 *A Good Speed to Virginia*, Robert Gray argued that God gave the Earth 'fee-simple' to man, yet most of the Earth is "possessed and wrongfully usurped by wild beasts, and unreasonable creatures, or by brutish savages." Gray assured the colonizers that they were actually *not* taking the land from anyone: those 'savages' "have no interest in it, because they participate rather in the nature of beasts than men." To be human is to make the Earth productive. It is to be a busy 'Bee,' Gray writes, that flies abroad "to gather the pleasures and riches of the earth."

In his 1689 *Two Treatises of Government*, Locke gave this theory of being human a fuller articulation. The central theme of Locke's political philosophy is *increase*. This can be understood as 'P' for *productivity* in White's formula for progress: E X T = P. Recall, the goal is *increasing production* to make a bigger factor M (for means) to provide not just security but also *convenience*. Locke said that we are instructed to be fruitful and to make more of "the conveniences of life." But nature is miserly. The natural world that God gave to humanity is "always one and the same, and it does not grow in proportion to the number of people born… Whenever either the desire or the need of possession increases among men, there is no extension, then and there, of the world's limits" (in Goldwin 1987, p. 494).

Nature never grows by one inch, even when our desires and needs do! This results in a zero-sum game. Whatever one person gains is only at the expense of taking from someone else. It is not within nature's power to extend her limits, so how can there be an improvement in the condition of humans generally? How can we *increase*?

The answer: human labor. What is beyond the power of nature (to increase) is well within the power of humans. Indeed, land without labor is scarcely worth anything, because "labor makes the far greatest part of the value of things we enjoy in this world" (ch. 5, sec. 40). Think back to the old ranch in Denton that is now home to a new shopping center complete with a Whataburger and so much more. The ranch was originally owned by James Newton Rayzor, who had arrived in Texas from Kentucky in 1866 as a child in an ox-drawn wagon. He amassed considerable wealth from his Alliance Milling Co. Their flour, brand-named "Peacemaker," won first place at the Texas State Fair so many times that it was eventually barred from competition.

Sometime in the 1920s, James bought a parcel of land to the northwest of the old courthouse on the downtown square and named it Rayzor Ranch. An old newspaper story reports that James' son Jesse managed a herd of Texas longhorns on this family ranch in the 1960s. Fencing and grazing that land vastly increased its productivity. Locke wrote that "the provisions serving to the support of human life produced by one acre of enclosed and cultivated land are…ten times more than those which are yielded by an acre of land of an equal richness lying waste in common." He who encloses land produces "a greater plenty of the conveniences of life from ten acres than he could have from a hundred left to nature, [and] may truly be said to give ninety acres to mankind" (ch. 5, sec. 37).

Now that Rayzor Ranch has been up-valued into shopping and dining, it is producing hundreds of times more conveniences and wealth. Tax revenues are through the roof. Imagine the difference in value of those

1500 square feet of land before and after the Whataburger. For Locke, it is the labor that went first into the ranch and then into building and operating the restaurants and shops that accounts for all the value. Since that is the case, that land can rightfully be claimed as private property. The creation of value via labor, property, and capital is simply development or progress.

The Native Americans like the Caddo and Comanche that once lived on the land, now known as Rayzor Ranch Market Place, had a super abundance of natural resources. But those are only resources in potential. Realizing that potential requires the energy of human hands. As a result, Locke writes that they were "rich in land and poor in all the comforts of life…for want of improving it by labor have not one-hundreth part of the conveniences we enjoy" (ch. 5, sec. 41). Locke notes that a king in America is worse off than a day laborer in England, because the poorest members of a productive society reap material benefits unattainable by even the elite of unproductive societies.

The "industrial and rational" few who successfully harness the power of *increase* begin to accumulate wealth. Their holdings grow too large for them to secure on their own. The government of civil society, Locke argues, is created to protect these possessions from the "fancy or covetousness of the quarrelsome and contentious." The art of government is "the increase of lands." In other words, the job of the public sphere is to manage, secure, and cultivate the growing economy.

Indeed, this *increase* unlocked by labor is what we call economic growth. The energy of human labor—our exo-somatic metabolism—is the economy. As the economy expands and becomes more sophisticated, it becomes clear that we don't so much "gather the pleasures and riches of the earth" as Gray wrote, rather they convene for us. Each of us doing our small, specialized part contributes to an unfathomable whole that makes commodities more and more comfortably and securely available. There are many virtues or skills involved here, but the beauty of this formula for *increase* is that it is primarily driven by selfish motives and narrow intentions.

Bernard Mandeville's 1714 *Fable of the Bees* captures this well. His poem tells how the busy-bee machinations of private vice give rise to the luxuries and conveniences of a prosperous society. Though we might grumble about the decadence and waste of a consumer culture, beware any proposal that actually seeks to restore a program of self-control. A truly virtuous hive would decline into poverty. Greed is good! It's the only motive strong enough to get people to do the hard labor necessary to create the easy life.

We should be deeply suspicious of any appeals to the *virtue* of limiting desire. This is an age-old strategy used by the elite to instill guilt. The masses limit themselves to their lowly stations while those on the top party on and laugh at the fools duped into believing in the 'virtues.' Mandeville writes that the "Wise Men" sought to make "the People" believe "that it was more beneficial for everybody to conquer than indulge his Appetites, and much better to mind the Publick than what seem'd his private Interest." The elite decided to call "Vice" anything that is self-interested and "Virtue" anything that is self-denying.

This is what Nietzsche would later call the creation of a slave morality by the religious elite who promise bogus otherworldly rewards. For example, Proverbs 16:32 claims, "He who conquers himself is stronger than he who conquers the strongest cities." But that is a lie designed to hide the real strength! The point of Mandeville's fable is to show that "the most elegant Comforts of Life" can only come from "an industrious, wealthy and powerful Nation." True wisdom involves embracing the "beautiful Machine" that arises from the "contemptible Branches" of self-interest. That 'Machine' is the market economy that is able to *increase* once civil society gives the energies of labor the protection of private property.

Locke's formula gives us the best metrics to track progress: income and economic growth. The economy is just the conversion of energy and money in and out of different forms. And energy and money are both simply means, parts of factor M, that can be used however one chooses. By accessing more energy to grow the economy, we are providing the means for people to answer the question of the good life as they see fit. Indeed, we are simply growing an ever-more easily accessible buffet of choices for them to consider. We are boosting individual capabilities. As the philosopher of energy Alex Epstein (2014) notes, money may not buy happiness, but it "gives us *resources* and therefore time and opportunity to pursue happiness" (p. 13).

* * *

It was Adam Smith who would take the ideas of Locke, Mandeville, and others and synthesize them into a theory of capitalism. Smith also doesn't figure prominently in books on energy, unless we count the obligatory nod given to him before extolling the virtues of the free market. Yet Smith was also thinking about human energies as they take shape in desire and labor.

Smith put the unlimited drive to amass wealth (the very moral psychology that Aristotle tried hard to banish) at the root of his philosophy. Actually, what we want, Smith argues in his 1776 *The Wealth of Nations*, is to "better our condition." It's just that most people take this to mean "an augmentation of fortune." The Invisible Hand, like a divine trickster, implants in us this crucial confusion of wealth with betterment. This is best explained in his 1759 *The Theory of Moral Sentiments* where he diagnoses the human capacity to be so distracted by the pursuit of means as to lose sight of the ends.

The "poor man's son," for example, who has been afflicted with ambition looks at the wealthy people around him and imagines that if he only had riches and servants like them he could be happy. So, "with unrelenting industry he labours night and day" to get the education he needs to get the job that will earn him those riches. Smith wryly notes:

> Through the whole of his life he pursues the idea of a certain artificial and elegant repose which he may never arrive at, for which he sacrifices a real tranquility that is at all times in his power, and which, if in the extremity of old age he should at last attain to it, he will find to be in no respect preferable to that humble security and contentment which he had abandoned for it. (Pt. IV, Chap. I)

Like hamsters on the wheel, we'll never get wherever we think we are going.

Admittedly, this sounds like an attack on the orthodoxy and the parable of the businessman, but Smith is a wily thinker! He goes on to argue that it is good that nature imposes "this deception" upon us, because it "rouses and keeps in continual motion the industry of mankind." The pursuit of wealth may leave the individual dissatisfied, but it is what prompted the cultivation of ground, the founding of cities, and the improvement of science and technology "which ennoble and embellish human life."

Smith paints a complex picture—he is rarely given credit for the many twists and turns of his thinking by those who wish to use him as a strawman. Echoing Locke, Smith argues that those few "lordly masters" who succeed in amassing great stocks of land and wealth cannot possibly consume all they have. Despite their "natural selfishness and rapacity," they wind up benefiting everyone. They only intend their own convenience and the gratification of "their own vain and insatiable desires." But in their maddening search for happiness they "are led by an invisible hand" to distribute the wealth to those they employ and to society more generally. As a result, everyone is equal with respect to "what constitutes the real happiness of human life," namely, "ease of body and peace of mind." Indeed, "the beggar, who suns himself by the side of the highway, possesses that security which kings are fighting for."

Once again, labor—the energy of our bodies and our hands guided by human intelligence—is the crux of the story. In the nascent capitalist economies of Smith's time, labor had begun to emerge from the private sphere of the home and the stifling bonds of the master-apprentice relationship. As labor became a social activity, its character could alter. In particular, it could be *divided* or specialized in ways that would never work in the home-spun, handcraft economies of old.

The chief means by which the capitalist (or lordly master) expands his wealth is the division of labor. Smith famously illustrates this with a "small manufactory" that makes pins. By dividing the task into its constituent parts (and assisting the tasks with machinery) the laborers could make 48,000 pins a day. If each laborer were to perform the task alone, the daily output would have been at most 200 and perhaps not even a single pin. Such massive gains in productivity from specialization are then rendered fruitful by the invisible hand of the market that convenes all the pieces into final commodities. Smith, like Hume, gives the passions a privileged place above reason, because the convening of a productive world just *happens* through the urges of all those poor man's sons and not through any rational plan.

Indeed, the invisible hand, which both drives and is driven by insatiable desire, does precisely what human reason is *incapable* of doing, namely, coordinating the kinds of conveniences like those found at Rayzor Ranch. No central planner could do that. To put this in a broader context, I began this book with a call for expanding our imagination to close the gap between what we are capable of doing and what we are capable of understanding. But the beauty of the invisible hand and the free market is that *we don't have to understand.* Individuals don't need to imagine or grasp the whole—that's impossible! The important moral imperative is to ensure that human creativity is free to be innovative.

Locke and Smith on one hand and Karl Marx on the other helped to spawn the great rival systems of modernity—capitalism and communism. Yet beneath that massive schism, there is the same centrality of labor. In many ways, Marx only codified what Locke had begun. For Marx, the human essence is labor. We are the *animal laborans.* And the laboring class is a bundle of potential energy. They are "labor power" or *Arbeitskraft,* which provides a surplus over and beyond mere metabolism and biological reproduction. This surplus energy generates an increasing flow of wealth. Its fertility is the same as that of natural processes where one man and one woman can produce by multiplication a great number of offspring.

Labor power is just another form of the "perfect similarity" of energy. For what is bought and sold on a market characterized by the division of labor is not individual skill or anything uniquely distinguishing about an individual. The skill of handcraft modes of making gives way to the more efficient modes of standardized production. The labor power used in productivity and bought and sold on labor markets is something possessed by everyone in approximately the same amount. It is a universal quantity that can be plugged into the social process of productivity.

In Locke's time, there were still crucial limits to this growing process. For it to truly accelerate, labor would have to be not just protected by government and not just divided, but also mechanized. Locke lived in a world powered largely by human and animal muscle and phytomass. The fossil economy was just starting to take hold. By the time Locke was writing, over 500,000 tons of coal were being shipped annually to London. He had what we would now probably call asthma, and the air pollution often made him sick. He went into self-imposed exile in Holland out of fear for his life—both at the hands of the king and of the smog. Dutch cities offered political and atmospheric safe harbor: they used peat for fuel, which burned cleaner than coal.

Despite the increasing use of coal, the prime mover capacity of Locke's world was hardly different from the Iron Age. Watt's steam engine wouldn't debut until 1776. Even by 1854, as Marx was settling into life in London, England's largest overshot waterwheel delivered just 600 horsepower. That was the pride of the most powerful nation on Earth, but it is the equivalent of just two of the bulldozers humming around Rayzor Ranch, which in turn is just one of the new shopping centers in north Texas. The largest dump trucks at the largest coal mines today produce 23,000 horsepower.

Slavery, as noted, was another way to increase productivity. Recall that Aristotle considered labor to be our enslavement to necessity. The only way to win some measure of freedom from this condition was to dominate others and force them into labor to provide the necessaries of life. This is why slavery forms a significant part of Aristotle's *Politics* (Barnes 1984). The mastery of slaves is the human way to conquer necessity, therefore, he argues, it is not against nature: "For that some should rule and others be ruled is a thing not only necessary, but expedient; from the hour of their birth, some are marked out for subjection, others for rule" (*Politics* 1254a20). In managing the household, Aristotle argued, certain instruments are needed. The slave is "an instrument of action" and "the servant is himself an instrument for instruments."

Then Aristotle mentions the tripods of Hephaestus, an early automobile with golden wheels, which according to Homer "could automatically enter the assembly of the gods and again return to their residence." Aristotle notes that if we possessed instrument such as these—that did labors automatically—then "chief workmen would not want servants, nor masters slaves" (1254a1). Yet he does not propose a political project to develop these kinds of instruments in the name of human liberation.

But that is precisely what the modern energy paradigm does! Yes, Locke was deeply involved in the slave trade. Yet he also helped to create a worldview and a world that would render human slavery obsolete. This is a far cry from Aristotle's smug acceptance of 'natural' slavery. Milton Friedman, an architect of the neoliberalism that would later build on Locke's liberalism, put it colorfully:

Industrial progress, mechanical improvement, all of the great wonders of the modern era have meant little to the wealthy. The rich in ancient Greece would have benefited hardly at all from modern plumbing – running servants replaced running water…the great achievements of western capitalism…have made available to the masses conveniences and amenities that were previously the exclusive prerogative of the rich and powerful. (Friedman and Friedman 1980, p. 148)

The moral clarity of this sentiment is beyond reproach. I will side with Jefferson (though he too was very much tangled up with slavery) when he wrote that "the mass of mankind has not been born with saddles on their backs, nor a favored few booted and spurred." Many nineteenth-century abolitionist arguments hinged on mechanization as a vehicle of liberation and human dignity. In 1891, Oscar Wilde wrote, "Unless there are slaves to do the ugly horrible uninteresting work, culture and contemplation become almost impossible. Human slavery is wrong, insecure and demoralizing. On mechanical slavery, the slavery of the machine, the future of the world depends."

Of course, Marx saw how human drudgery and exploitation accompanied the "colossal productive forces" unleashed by the capitalist "means of production" (factor M). But he did not blame *increase* itself or the harnessing of labor power. Capitalism, he believed, had been a necessary stage in history. Indeed, it was a crucial phase when humans developed the ability to command the energies of the world to create abundance. Capitalism created the conditions of material abundance that are necessary for freedom, but with its logics of private property and private ownership of the means of production it only allowed the few to enjoy freedom. Via organization of the laboring class, communism would distribute the fruits of industry more equitably.

In the mid-nineteenth century, Marx could see just how fruitful modern industrialism was becoming. By the 1880s, the world's steam engines were performing the labor equivalent of 3 billion humans on a planet with a population of only 1 billion. In 1940, Buckminster Fuller coined the term "energy slave" and argued that machines not only replaced human muscle power, but could perform tasks no human could do by working under extreme heat or pressure, for example, or performing to exacting standards of precision.

A single gallon of gas contains the energy equivalent of 490 hours of strenuous human labor, which would cost at least $6500 at minimum wage. Roughly 99% of the labor in contemporary society is done by fossil fuels (see Morgan 2013). By some estimates, the average American commands 186,000 "machine calories" daily (Epstein 2014). That's the equivalent of ninety-three humans, but without any slavery.

The story of modern energy is about human labor creating *increase* or productivity. The capital and value generated then offload hard labor onto the backs of machines while increasingly *convening* a world of commodities. It has been a hard struggle, and we would do well to remember all the energy—all the sweat—that has been poured into our world of Whataburger, air conditioning, and smooth road trips. So, I will conclude with an excerpt from the diary of James Newton Rayzor:

> Our fathers and mothers came to Denton County in the time of the buffalo and the wild Indians, and from a wilderness carved out the wonderful civilization we are now enjoying. Through hardships, privations, and perils, they laid well the foundations of our Baptist Zion, in which, in comfort and ease, we supinely rest.
>
> None but those who have passed through such testing times can appreciate how nobly they wrought. They blazed out the way for us, and have gone home to receive their imperishable reward, and have committed the task of building a worthy superstructure upon the foundation they have left us. Let us, guided by Divine wisdom and consecrated leadership, strive to prove ourselves worthy of the legacy they have bequeathed us.
>
> Should we fail to make the way plainer, smoother, straighter, brighter and better for those who shall come after us to walk in, then our living shall have been in vain.

Bibliography

Barnes, Jonathan, ed. 1984. *The Complete Works of Aristotle.* 2 vols. Princeton: Princeton University Press.

Epstein, Alex. 2014. *The Moral Case for Fossil Fuels.* New York: Penguin.

Friedman, Milton, and Rose Friedman. 1980. *Free to Choose: A Personal Statement.* New York: Harcourt Brace Jovanovich.

Goldwin, Robert. 1987. John Locke. In *History of Political Philosophy*, ed. Leo Strauss and Joseph Cropsey. Chicago: University of Chicago Press.

Gray, Robert. 1609. *A Good Speed to Virginia*. London: Felix Kyngston.

Hume, David. 1748. *An Enquiry Concerning Human Understanding*. Project Gutenberg. https://www.gutenberg.org/files/9662/9662-h/9662-h.htm

———. 1777 (1985). *Essays, Moral, Political and Literary*, ed. E.F. Miller. Indianaopolis, IN: Liberty Classics.

Locke, John. 1689 (1994). *Two Treatises of Government*, ed. Peter Laslett. Cambridge: Cambridge University Press.

Mandeville, Bernard. 1714. Fable of the Bees. *Project Gutenberg*. https://www.gutenberg.org/files/57260/57260-h/57260-h.htm

Mitcham, Carl. 2019. *Steps Toward a Philosophy of Engineering: Historico-Philosophical and Critical Essays*. London: Rowman & Littlefield International.

Morgan, Tim. 2013. Perfect Storm: Energy, Finance, and the End of Growth. *Tullett Prebon*, issue 9, https://www.tullettprebon.com/Documents/strategy-insights/TPSI_009_Perfect_Storm_009.pdf

Smith, Adam. 1759. *The Theory of Moral Sentiments*, ed. Knud Haakonsen. Cambridge: Cambridge University Press.

———. 1776 (2003). *The Wealth of Nations*. New York: Bantam Books.

Tierney, Thomas. 1993. *The Value of Convenience: A Genealogy of Technical Culture*. Albany: SUNY Press.

Tredgold, Thomas. 1828. *Civil Engineering*. Used in the Royal Charter of the Institution of Civil Engineers and published in *The Times*, London, June.

Wilde, Oscar. 1891. The Soul of Man under Socialism. *Project Gutenberg*. http://www.gutenberg.org/ebooks/1017

11

Decoupling

The solution to the unintended consequences of modernity is, and has always been, more modernity.
Michael Shellenberger and Ted Nordhaus *(2011)*

On the morning of December 5, 1952, a thick fog materialized over the streets of London. "Pea-soupers" like this formed through a combination of natural fog and smoke from the coal-fired industries and coal-heated homes of London. They were common affairs. Indeed, early movies like the 1938 film "The Divorce of Lady X" often incorporated smog into their plotlines. Alfred Hitchcock's 1927 "The Lodger" depicts a serial killer who operates under the cloak provided by the smog.

The smog of 1952, however, was unique. A temperature inversion locked the city's poisonous emissions in place, creating a 30-mile-wide swath of increasingly toxic air full of sulfur. Cars were abandoned on the road, their headlights unable to cut through the daytime blackness. Conductors walked in front of the iconic double-decker busses with flashlights attempting to light their way. Pedestrians wheezed through the streets, slipping and falling on the greasy black film coating the sidewalks. Their faces were streaked and their nostrils were packed with soot.

Children were sent home from school out of fear that if they waited any longer they would not be able to find their way home. Even the movies about the smog couldn't be watched, because the actual smog had yellowed the screens at movie theaters, making them unusable. By the time it was over, the Great Smog of London had exacted a gruesome toll, much worse than any serial killer. Current estimates put the total deaths between 8000 and 12,000 people.

Recall that Locke had to escape the London smog way back in the seventeenth century. Did the formula of *increase* that he helped to write act as a recipe for disaster? Well, if the story ended in 1952, we might have to say 'yes.' We would have to abandon our "orthodox faith," the modern energy paradigm.

But the story doesn't end there. In 1956, Parliament passed the Clean Air Act. London continued to pioneer zoning laws to protect public health and safety. And technologies advanced to burn coal more cleanly and efficiently and to replace coal with less-polluting energy sources. In 1952, there were 200 micrograms/m^3 of suspended particulate matter (SPM) in London's air, and the GDP per capita was about $11,000. In 2015, those figures were 15 micrograms/m^3 and over $38,000 in constant dollars (Ritchie 2017). That's 3x the prosperity with 13x *less* air pollution. More economic growth—more of Lockes' *increase*—with fewer environmental and health harms.

That is the process of changing worse problems (e.g., exposure to the cold, squalor, and disease of medieval London—remember the plague!) for first world problems (smog) that we continue to mitigate through the application of human intelligence. We can now give it a name: decoupling. At first, economic growth comes hitched to environmental harms—you can't have the former without the latter. But then we reach a point of "peak impact" and the environmental harms decrease even as growth continues. In other words, we can have our cake and eat it too. We can have more prosperity *and* a healthy environment with a stable climate.

Admittedly, the modern energy paradigm appears to be unsustainable. After all, it is premised on the rejection of the *limits* built into the pre-modern worldview. But there is another kind of logic built into the modern energy paradigm. As wealth accumulates and our world becomes ever-more secure, we pass the point of peak impact on the planet. Indeed, the modern energy paradigm is not blind to threshold effects. It is premised on the most important one! London, for example, crossed a threshold as early as 1870. That's when its SPM emissions peaked. Ever since then, its story has been about declining pollution and increasing wealth. This kind of development pattern is common, and is frequently graphed on an environmental Kuznet's Curve similar to Fig. 11.1.

That will be the story for the whole planet and all of modern civilization. Yes, as with London, there will be casualties along the way. But we have to keep the big picture and the long-term trends in mind. When we look back over the past two hundred years, we find the same story again and again of initially harmful growth becoming progressively less and less impactful. What happened in London will happen in Beijing and New Delhi. This isn't 'faith,' it is just extrapolating well-established trends that come about through the logic of modernity.

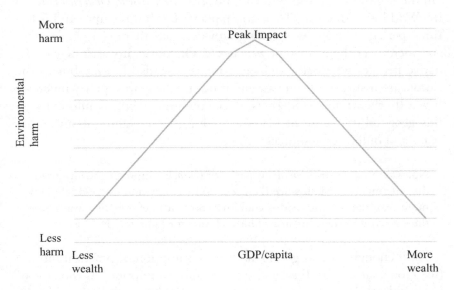

Fig. 11.1 Decoupling

To freeze progress now out of fear of the risks of technology is to fall victim to an irrational risk-assessment. This is understandable, given how much the media loves to prioritize the click bait of negative news and disasters. But this is incredibly dangerous. Recall the millions of people living in energy poverty and exposed to the perennial wickedness of the Fates. The risks of modernization are real, but they must be assessed from within this frame of reference. We need growth, it just needs to be *green* growth. This is the conclusion of many of the world's most important development organizations, including the European Commission (2013), the Organisation for Economic Co-operation and Development (OECD) (2011), the United Nations Environment Programme (2014), and the World Bank (2012).

Even though we live on a planet with an additional 7 billion people, Locke's formula is still the right one. Let's unpack it and put it into our contemporary context.

*   *   *

"In the beginning," Locke wrote in the *Second Treatise of Government*, "all the World was America." Translation: planet Earth was empty. In Locke's time, perhaps Europe was getting crowded, but there is a whole other continent over there, a whole *new world*. Of course, it wasn't new to the native peoples who had lived there for thousands of years, but to the European, it sure did seem vast and if not totally empty, pretty much so. It was (to them) pure wilderness. Locke built his political philosophy on this geographical sense of place—this intuitive certainty about empty space and unlimited resources:

> Nor was this *appropriation* of any parcel of *Land,* by improving it, any prejudice to any other Man, since there was still enough, and as good left; and more than the yet unprovided could use. So that in effect, there was never the less left for others because of his inclosure for himself. (ch. 5, sec. 33)

"Still enough, and as good left" is an assumption of practical infinity (see Frodeman 2019). It seemed a reasonable assumption. At the time of Locke's death in 1704 there were perhaps 250,000 people living in the colonies, less than half the population of London at the time (the native population in what is now the United States totaled perhaps 5 million, though estimates vary widely). The US population grew to 5 million in 1800, 76 million in 1900, and 281 million in 2000. It is now 330 million.

By the 1960s, this growth sparked a new environmental consciousness, one that had a metaphysical dilemma. It sought to defend a sense of limit, a sense of the proper human reach, but within the modern scientific worldview that had eliminated the natural grounds for such limits. Modern science offers no sense of proportion, no normative claims about sufficiency or propriety. Unable to appeal to *telos* and *energeia*, modern environmentalism turned to the new science of ecology and related sciences of demography. Its ruling metaphors became the footprint (the carbon footprint for climate change) and the petri dish. Like bacteria in a petri dish, we will eventually hit the wall and population will collapse. The reigning principle became *precaution*—beware tinkering with powers you don't understand. Things will get *out of control*.

The patron saint for this new science of limits was the English scholar Thomas Robert Malthus, and its holy text was his 1798 *An Essay on the Principle of Population*. We are bound, he argued, to get caught in the "Malthusian trap," where increased productivity leads to increased population that in turn leads to increased consumption and finally a shortage of resources and a population crash. Malthus was an Anglican cleric and saw in this a divine lesson to teach moral restraint and virtue. He opposed the Poor Laws, which were designed to improve the lot of the proletariat or laboring class, because they would only artificially inflate population leading to more suffering and misery when we inevitably hit the wall.

In the 1960s and 1970s, leading scientists embraced this Malthusian image of a strictly limited natural world harshly constraining the ambitions of human *increase*. They too extrapolated an ethics from the science. As noted earlier, the ecologist Garret Hardin (1975) argued that we need to adopt a "lifeboat ethics," where rich nations only help poor populations at the expense of dooming everyone. The Earth, like a lifeboat, has a strictly limited carrying capacity. This too was the message of the influential biologist Paul Ehrlich in his best-selling 1968 book *The Population Bomb*. Call it sinking or call it exploding: Locke's logic of an unfettered E leads to doom.

But no, it doesn't! And everyone seemed to *know* that even if they professed to *believe* Ehrlich when he talked about doom and gloom on *The Tonight Show with Johnny Carson*. After all, throughout the mid- and late twentieth century, an unprecedented boon in economic growth was unleashed. If everyone really believed Ehrlich, why didn't they scale back and consume less? Why didn't the hippies win the culture wars? It's because we don't have to constrain ourselves! Indeed, what we need to do is empower ever-more minds with ever-more knowledge.

There were a handful of economists and other intellectuals who understood and had explicitly theorized what everyone else implicitly believed. They worked at institutions like Resources for the Future and the American Enterprise Institute. One of their leaders was Simon Kuznets, who started tracking the Kuznet's Curves of development mentioned above. Another leading voice was the economist Julian Simon. In a famous wager, Simon asked Ehrlich to choose any five commodity metals. Simon bet that all five would be lower in price after ten years. Ehrlich was sure their prices would increase. After all, if we spend ten years mining tin there will be less of it, thus it will be rarer and cost more. Yet Simon won the bet: all five metals were cheaper despite ten years of rampant consumption.

Simon won because he was using Locke's winning formula. Recall: nature is not valuable, human labor is. So-called natural resources are really the work of human ingenuity. It's not like aluminum foil grows on trees! We don't live on the Big Rock Candy Mountain with lemonade springs. We have to deal with scarcity. But scarcity is just another name for an increase in price. When prices go up on a free market, entrepreneurs and engineers have an incentive to find more of a resource or a replacement for it. If copper prices go up, we can invent fiber optic cables. After all, people don't want copper, they want the telecommunication services. If natural gas and oil prices go up, we can invent hydraulic fracturing, 3-D seismic imaging, and horizontal drilling to tap into previously inaccessible reservoirs. In this way, proven reserves of fossil fuels actually *increase* even as we consume ever more.

As just one example, in 2019 a group of industry experts once again revised estimates of US natural gas supplies upward (Anchondo 2019). It turns out that there is 20% more producible natural gas than they thought just two years earlier! We keep getting better at finding and recovering energy. Before Simon, the Texas economist Erich Zimmerman already drew the logical conclusion: since we essentially make resources, conservation "to protect the interest of future generations [is] unnecessary" (in Nikiforuk 2012, p. 141). Technology will find replacements and the human economy can keep levitating and expanding like a helium balloon. As Simon's acolytes and energy analysts Robert Bradley and Richard Fulmer (2004) argue, "The lesson of history is that in free societies, individuals produce more energy than they consume." Capitalism is a perpetual motion machine.

Locke's image of infinity was right. The Earth is not some cookie jar with a limited number of resources. The human mind is, Simon wrote, *The Ultimate Resource* (1981). Resources spring from our minds in the same way that Athena sprang from the forehead of Zeus. When smart people succeed in finding more materials, making things more efficient, or inventing replacements, prices go back down and prosperity ratchets up. The environment gets cleaner as people get richer. Human ingenuity ("the work of our hands") hitched to the free market can solve any problem. Government can play a limited but crucial role here. Parliament's Clean Air Act, for example, incentivized clean tech innovation. Through government mandates, my washing machine is more energy efficient than my mother's was.

Forget Aristotle's *dynamis* of character, what we need is the *dynamism* of the knowledge economy! From Simon and his later protégés (e.g., Lomborg 2001), we can distill key elements of the winning formula: (a) as we use more resources and grow the economy, we initially harm the environment; but (b) we then pass the threshold of peak impact and the environment actually improves (the wealthy have the luxury of caring for nature); (c) doomsayers like Ehrlich err in not considering all the benefits of technology and all the risks of regulations; (d) nature is almost always more dangerous than technology; (e) the media exaggerate bad news because it sells; (f) the ultimate standard for measuring progress is *human flourishing*; (g) the best metric for human flourishing is economic growth; and (h) the global economy is growing. Ergo: progress.

Locke's modernity, in other words, is also the key to an ecomodernity. Ecomodern thinkers like Ted Nordhaus and Michael Shellenberger understand well that the solutions to the problems caused by technology is more and better technology. We solved the ozone hole crisis not by giving up refrigerators or air conditioners, but by inventing better ones. This too is how we will solve the so-called climate crisis.

*   *   *

The Malthusian worldview of finitude has long jousted with the modern energy paradigm, even at the highest levels of government. In 1926, the well-known economist John Ise said that the automobile represented extravagance and theft from future generations: "We probably have the right to prefer our thousandth joyride to the thousandth joyride of our grandchildren, but whether we have the right to deprive them of their only ride in order that we may indulge ourselves with two thousand such rides is another question" (in Nikiforuk 2012).

Ise suggested a threshold for appropriate consumption (neither too little nor too much) on the basis of an assumed finitude that compels a zero-sum tradeoff. Whatever we consume today won't be available tomorrow. But this is the worldview that Locke already thought his way out of in the 1670s! Through our ingenious labors, we will develop more resources in the future to replace consumption in the present. This is a major reason why the orthodoxy isn't nearly as *judgmental* as Aristotle with all his talk about the *good*. If we can all have whatever we want without compelling others to make sacrifices, then why bother wading into contentious debates about the good life? Everyone can do as they please.

Jimmy Carter was among the last politicians to speak of energy in terms of character and virtue. Like Ise, he did so on the assumption of finite resources. When Carter took office in 1977, the energy crisis dominated the political agenda. In response, he gave a series of speeches on energy, including one in which he wore a sweater to encourage Americans to conserve energy by turning down their thermostats. In stark terms, Carter informed the nation that it was running out of oil and gas. To respond appropriately, "we must not be selfish or timid." It would entail a massive effort that would be "the moral equivalent of war."

Carter's energy policy consisted of ten principles, with *limit* at its core: "The sixth principle, and the cornerstone of our policy, is to reduce demand through conservation." This will, he continued, "demand that we make sacrifices and changes in every life. To some degree, the sacrifices will be painful – but so is any meaningful sacrifice." In 1979, he gave a speech titled "Crisis of Confidence," though it has since come to be known as the 'malaise' speech even though he never used that term.

Though Carter had been an engineer trained in nuclear physics, he reverted back to Aristotelian science. He framed the energy challenge as a matter of moral character more than technology: "Too many of us now tend to worship self-indulgence and consumption. Human identity is no longer defined by what one does, but by what one owns." It was reminiscent of an 1886 speech by Theodore Roosevelt: "We must keep steadfastly in mind that no people were ever yet benefited by riches if their prosperity corrupted their virtue" (in Rao 2018).

Carter's speech was well-received at first, but quickly became a joke. He was painted as the weak-willed national scold. Indeed, the Moral Equivalent of War speech came to be known by the catty acronym: MEOW. Carter was proposing some sense of limit. And he ran into the perennial problem of how to define the parameters for a doctrine of the mean and especially how to impose this as a matter of state policy. There is no non-arbitrary way to draw a line for the upper-end of the threshold. When do we have enough? And what gives Carter the authority to dictate lifestyle choices? He's the president, not the pope!

At the 2018 CERAWeek energy conference, US Energy Secretary Rick Perry lamented a "fundamentally flawed energy policy" of the 1970s. He was talking not just about Carter but also the predominant worldview about limits, which was widely assumed to be a clear-eyed realism: the Earth is finite, after all, so our high-energy party has got to end soon. Perry said that thankfully a "new energy realism" had since risen to the fore (Perry 2018). There was no shortage of energy in the 1970s, only a "shortage of imagination and a loss of confidence in our ability to innovate." Of course, this 'new' realism was as old as Bacon, Locke, and Descartes.

With science and technology, Perry continued, we were able to "transcend limitations" and usher in an age of affordable energy abundance.

Better yet, even as the economy grew, the environment improved. Innovation, not regulation—growth, not limits—is the path to prosperity *and* sustainability. This means "we don't have to tolerate the intolerable…we don't have to accept hideous sacrifices that harm the poorest among us."

Bill Gates, the world's second wealthiest man, has quickly become a major prophet of the energy orthodoxy with his Breakthrough Energy Coalition (launched in 2015) and Breakthrough Energy Venture (launched in 2016). Together, the Breakthrough Energy initiatives are pumping over $1 billion into cutting-edge companies and technologies "to make sure that everyone on the planet can enjoy a good standard of living, including basic electricity, healthy food, comfortable buildings, and convenient transportation, without contributing to climate change."

The homepage of Breakthrough Energy notes that by the middle of the century "the world will use twice as much energy as we do today." The website continues:

> And that's a great thing: The more access to energy people have, the larger our economies grow and the better our lives become. But to get there, we need different tools than the ones that have served us in the past.

It's instrumental ethics again. We know where to go (bigger!), we just need to figure out new tools for getting us there in a way that doesn't contribute to climate change. Gates favors nuclear power. He is also very enthusiastic about artificial photosynthesis to create hydrogen fuel. In such ways, we will reach peak impact and decouple the growing human world from the natural environments of planet Earth.

Both Gates and the philosopher of energy Alex Epstein see modern energy services as giving us superpowers. Imagine, Epstein writes, if Superman used his powers to help developing nations. He could "melt iron, forge steel, plow fields, build buildings, even run an electrical system…*He could transform any place for the better.*" But that's just exactly what we do! "Using human ingenuity, we have made ourselves into supermen" (Epstein 2014, p. 41).

Bill Gates and Melinda Gates (2016) also describes energy as a 'superpower' that enables us to control the world around us. He writes:

If I had to sum up history in one sentence it would be: 'Life gets better—not for everyone all the time, but for most people most of the time.' And the reason is energy. For thousands of years, people burned wood for fuel. Their lives were, by and large, short and hard. But when we started using coal in the 1800s, life started getting better a lot faster. Pretty soon we had lights, refrigerators, skyscrapers, elevators, air conditioning, cars, planes, and all the other things that make up modern life…Without access to energy, the poor are stuck in the dark, denied all of these benefits and opportunities that come with power.

Like Gates, Epstein argues that life improves as we command greater amounts of energy. Life expectancy, GDP per capita, and access to clean water go up, infant mortality and climate-related deaths go down. Cheap, reliable, and plentiful energy allows for the construction of robust and resilient technological cultures capable of shielding us from natural dangers.

Epstein calls energy "machine calories." All of our machines, from incubators in hospitals to jet aircraft, need a constant input of calories. As for humans: "We desperately need machines to do work for us, because we are naturally *very weak*. Without machines to help us, we don't have anywhere near all the energy that we need to survive and flourish" (Epstein 2014, p. 40). The more energy we use per unit time, the more *power* we have. For "99% of human history," people "lacked power" and thus "spent their lives engaging in grueling physical labor just to keep their bodies going long enough to engage in the next day of grueling physical labor" (p. 41).

This is all correct, but Epstein's choice of superhero is misplaced. What he is describing is not Superman, whose powers are naturally gifted. Epstein, rather, is describing something closer to Batman, whose powers are the product of his ingenuity and technologies. Yet even Batman is not quite right as he is a self-made man. By contrast, we do not make the world around us in our own bat caves. Rather, it *convenes* for us via networks of innovation and supply chains. Unlike Batman, we don't have to make (or even understand!) our own power-boosting and fate-denying exoskeleton that we call "the grid."

As for climate change, Gates (2016) has an equation to capture the ultimate decoupling challenge of our age: decoupling global economic growth from carbon emissions. Actually, Gates' equation is not his own original work. In the 1980s, the Japanese scientist Yoichi Kaya developed a vital formula for climate models and climate policy analyses (see Pielke 2010). Known as the Kaya Identity, the formula encompasses all the tools in the policy toolbox for climate mitigation. It shows that there are four levers for stabilizing concentrations of greenhouse gasses in the atmosphere: population, per capita wealth, energy intensity of the economy (energy consumption per unit of GDP), and carbon intensity of energy (carbon dioxide emissions per unit of energy consumption).

Here is how Gates puts his related formula:

$$P \times S \times E \times C = CO_2$$

P stands for population, S for services, E for the energy required to provide those services, and C for the carbon produced by that energy. Population is going up. Energy is also going up and, he argues, should continue to do so, because modern energy services are the key to flourishing lives. The 'S' too "should go up." This "is everything: food, clothing, heat, houses, cars, TV, toothbrushes, Elmo dolls, Taylor Swift albums, etc." In other words, Gates' 'S' is the 'P' for production in White's formula about cultural progress—it is factor M. Again, the ultimate goal is to increase production to provide the good life of security and convenience.

That leaves us with just one pinch point: letter C for carbon. Gates concludes that the *only way* to get CO_2 down to zero is to get C to zero—we must develop zero-carbon energy technologies. This is a massive challenge, this next grand energy transition. But have faith! We have the recipe for success in hand. We just need the *willpower* to keep applying it.

Bibliography

Anchondo, Carlos. 2019. US Gas Supply 20% Higher than Previously Thought – Report. *E&E News*, September 11. https://www.eenews.net/stories/1061112181

Bradley, Robert, and Richard Fulmer. 2004. *Energy: The Master Resource*. Dubuque: Kendall Hunt Publishing.

Ehrlich, Paul. 1968. *The Population Bomb*. New York: Ballantine Books.

Epstein, Alex. 2014. *The Moral Case for Fossil Fuels*. New York: Penguin.

European Commission. 2013. *Living Well, within the Limits of our Planet: 7th EAP – The New General Union Environment Action Programme to 2020*. https://doi.org/10.2779/57220.

Frodeman, Robert. 2019. *Transhumanism, Nature, and the Ends of Science*. New York: Routledge.

Gates, Bill, and Melinda Gates. 2016. Two Superpowers We Wish We Had. *GatesNotes*, February 22. https://www.gatesnotes.com/2016-Annual-Letter

Hardin, Garrett. 1975. Lifeboat Ethics. *Hastings Center Report* 5 (1): 4.

Locke, John. 1689 (1994). *Two Treatises of Government*, ed. Peter Laslett. Cambridge: Cambridge University Press.

Lomborg, Bjorn. 2001. *The Skeptical Environmentalist: Measuring the Real State of the World*. Cambridge: Cambridge University Press.

Malthus, Robert. 1798. An Essay on the Principle of Population. *Project Gutenberg*. http://www.gutenberg.org/ebooks/4239

Nikiforuk, Andrew. 2012. *The Energy of Slaves: Oil and the New Servitude*. Vancouver: Greystone Books.

OECD 2011. *Towards Green Growth: A Summary for Policymakers*. May. https://www.oecd.org/greengrowth/48012345.pdf

Perry, Rick. 2018. The New Energy Realism. *CERAWeek*, March 7. https://www.energy.gov/articles/new-energy-realism-secretary-perry-remarks-cera-week-prepared-delivery

Pielke, Roger. 2010. *The Climate Fix: What Scientists and Politicians Won't Tell You About Global Warming*. New York: Basic Books.

Rao, Maya. 2018. *Great American Outpost: Dreamers, Mavericks, and the Making of an Oil Frontier*. New York: Hachette Books.

Ritchie, Hannah. 2017. What the History of London's Air Pollution Can tell Us about the Future of Today's Growing Megacities. *Our World in Data*, June 20. https://ourworldindata.org/london-air-pollution

Shellenberger, Michael, and Ted Nordhaus. 2011. Evolve. *Orion Magazine*, August 25. https://orionmagazine.org/article/evolve/

Simon, Julian. 1981. *The Ultimate Resource*. Princeton: Princeton University Press.

UNEP. 2014. *Decoupling 2: Technologies, Opportunities and Policy Options: A Report of the Working Group on Decoupling to the International Resource Panel*. UNEP. https://www.resourcepanel.org/sites/default/files/documents/document/media/-decoupling_2_technologies_opportunities_and_policy_options-2014irp_decoupling_2_report-1.pdf

World Bank. 2012. *Inclusive Green Growth: The Pathway to Sustainable Development*. International Bank for Reconstruction and development. https://doi.org/10.1596/978-0-8213-9551-6.

12

Prometheus 2.0

There can be no question of which was the greatest era for culture; the answer has to be today, until it is superseded by tomorrow
Steven Pinker (*2018*)

I began my defense of the modern energy paradigm (the so-called orthodox faith) with a parable. I then turned to a history of ideas and machines to demonstrate the logic of that story. It is a story about scientific knowledge and technological control and the replacement of worse problems by better ones. The mastery of energy reduces risks, liberates us from toils, and enriches our lives. Let's end with another parable. This one is the ultimate energy myth: the story of Prometheus and how he stole fire from the gods. It is an instructive myth but, as we'll see, it needs updating.

In the *Protagoras*, Plato has Protagoras give a "great speech" about how virtue can be taught. Protagoras turns to myth (see 320c-328d): "Once upon a time there were gods only, and no mortal creatures." When the time came, the gods fashioned the various creatures out of the elements of the Earth. They then ordered Prometheus and Epimetheus "to equip them, and to distribute to them their proper qualities." Epimetheus

© The Author(s) 2021
A. Briggle, *Thinking Through Climate Change*, Palgrave Studies in the Future of Humanity and its Successors, https://doi.org/10.1007/978-3-030-53587-2_12

(meaning 'hindsight') said that he would do the distributing and that his brother Prometheus (meaning 'foresight') would do the inspecting. It was agreed.

Epimetheus did a fine job equipping all the animals with what they need to survive. He gave some strength without swiftness, while he equipped the weaker with swiftness. Some he armed and some he gave other means of protection. For example, some had sheer size as their defense, others could burrow and others could fly. It was all designed to balance out, "with the view of preventing any race from becoming extinct."

But Epimetheus, "who not being very wise," left humans "naked and shoeless." Prometheus discovered this with no time to spare as the "appointed hour was approaching when man in his turn was to go forth into the light of day." Not knowing how to equip humanity, Prometheus stole the mechanical arts from Hephaestus and Athena. He also stole fire, because the mechanical arts (tools or technologies) "could neither have been acquired nor used without fire."

"And in this way," the story continues, "man was supplied with the means of life." Humans made clothing, built homes and cities, and otherwise "drew sustenance from the Earth" by using their command of fire, that is, energy and technology. Thus, in the mythic time-before-time, humans were gifted the divine spark of 'fire,' which should be understood as both our intelligence and our divinely sanctioned quest to use that fire to control the world around us.

With fire and technique, humans "had the wisdom necessary to the support of life, but political wisdom they had not; for that was in the keeping of Zeus, and the power of Prometheus did not extend to entering into the citadel of heaven, where Zeus dwelt." With the technical powers or virtues (this "share of divine attributes"), humans began to build houses, make clothing, and gather into cities. But without the political virtues (that other divine attribute), they constantly fell into war. Fearing that humankind would self-destruct, Zeus sent Hermes to bring justice (*dike*) and respect, modesty, or humility (*aidos*).

Hermes asked if he should distribute these virtues to all or just the few. Zeus replied, "To all. I should like them all to have a share; for cities cannot exist if a few only share in the virtues." When it comes to the mechanical arts, only the few (the experts) share in deliberations. But when it comes to the moral and political contexts of the mechanical arts, "every man ought to share in this sort of virtue." Collectively, the *polis* sets the boundaries for the technical arts so that they do not overreach or grow out of balance and lead to ruin. In short, *aidos* or 'proportion' is the humbling check on *hubris*. And *dike*, justice, is the governor to reign in the potential excesses of technical powers.

But proportion is precisely what modernity has rejected. Indeed, its prophets saw that virtue as a collective project of restraint led to intolerance and injustice. Justice comes about through freedom, the absence of constraints. This means that government needs to be subordinate to innovation and the process of economic growth that it fuels. It is this *growth* that lifts people out of the ultimate injustices imposed by the Fates. Even massive environmental projects today such as the American Green New Deal or China's Ecological Civilization are techniques for sustaining economic growth, not limiting it; for expanding and switching energy technologies, not curtailing energy flows.

The Promethean story as told in *Protagoras* is about liberation and transgression. Protagoras builds a governor into the tale by having Hermes distribute a tempering sense of shared political virtue. This entails training the wicked soul horse, but we can and must do without this part of the myth, especially in a globalized, multi-cultural world. The globe is not the ancient Greek *polis*. We do not share an ideal of the proper or excellent human being. Any limit imposed by that ideal form or *telos* would be an arbitrary and unjust imposition on people's own authentic self-development.

Our virtues are indeed the infinite. Progress is without limit. Recall the businessman in our opening parable. As an individual, he may rest, but as a symbol of the modern world, his logic is restless. We can always further elaborate the means, develop the technologies, grow the economy, secure the grid, and harvest more energy. A sense of limit would have to be supplied from some other story or source of meaning. Maybe you want to restrict your meat consumption or commute on your bicycle out of a personal spiritual or religious conviction. That's fine on an individual level, but we cannot justly build a society around restrictions on liberty. As John Stuart Mill (1861) wrote, "all honour to those who can give up for themselves the personal enjoyment of life…He may be a rousing proof of what men *can* do, but surely not an example of what they *should* do."

In her own parable, "Pyrrhus and Cineas," the existentialist philosopher Simone de Beauvoir (2004) puts things more forcefully. Pyrrhus was a Greek general and statesman. When his friend Cineas asks what his plans are after conquering the next city, Pyrrhus replies "conquer another one." And then? "Another one." Eventually, though, Pyrrhus stops the infinite regress by saying that he will 'rest.' Yet De Beauvoir is correct to argue that this is just a failure of his imagination. There is always another conquest to be had! Mars, here we come.

* * *

Let's call the version of the myth told by Protagoras Prometheus 1.0. It is a cautionary tale about how we need virtue understood as proportioning limits to counter-balance the powers that Prometheus gave to us. We need self-control as much as we need to control fate. For daring to steal fire, Prometheus is doomed to eternal punishment. He is bound to a pillar where an eagle devours his liver every day. His liver regrows every night so that the horror and torment continue perpetually. Similarly, Adam and Eve are doomed to ceaseless toil and labor when evicted from the Garden of Eden for eating from the Tree of the Knowledge of Good and Evil. These banishments are both symbols of that primordial energetic imperative—the incessant need to stoke the fires of metabolism. The liver grows and re-grows, Cain and Able must till and shepherd a new crop and a new herd every season. We must constantly re-assemble our bodies anew.

Both of these myths are based in the pre-modern energy paradigm, because they are about humans being put in their *proper place* by limiting their reach. Adam and Eve are given limits. Prometheus was unable to enter into the citadel of Zeus. We can understand that citadel as the house of knowledge. So, the proportionality or humility built into Prometheus 1.0 rests on a metaphysical claim: we must limit the project of attempting to control fate, because our ability to cognitively access reality is limited. There are ontological uncertainties—mysteries built into Being that we cannot know and, thus, cannot control. Wisdom counsels us not to overreach.

Prometheus 2.0, our modern story, is the rejection of this metaphysical limit. Once again, Descartes leads the way. True, his third maxim in *Discourse on the Method* is "to try always to master myself rather than fortune, and change my desires rather than changing how things stand in the world." But, don't be deceived! He is writing in dangerous times when the Church is still imposing the old paradigm of limits. As Leo Strauss (1952) argues, we have to learn how *to read between the lines*. Descartes' real project is precisely about controlling fate. He argues that we can follow a method to gain clear and distinct ideas that can establish an absolutely firm foundation on which to build our world.

We can overcome ontological uncertainty and achieve "epistemological certainty." This is to enter into the "citadel of heaven" where Prometheus 1.0 could not, or dared not, tread. We should take the Bible at its word: "by the sweat of your brow you will eat your bread." The sweat of our brow is caused by the heat of our intellect, our divine fire of creative innovation. Indeed we will rebuild Eden by fully realizing our divine potential energies!

The accelerationists and transhumanists are the latest prophets of the modern energy paradigm. For example, in his own essay on "Prometheanism," Ray Brassier (2014) shows how epistemological certainty is built into the winning formula that we have been discussing. It is the claim that we can overcome any boundaries to our knowledge and thus that we can overcome any limits and control any fate. Aging and death, most profoundly, are recast as solvable problems rather than "certain certainties" about the human condition. Epistemological certainty simply means that everything is amenable to a technological fix. All problems are solvable.

We can now see the future of the modern energy paradigm. The businessman has a very long way to go on the path of perpetual progress. Yes, he has overcome many limitations, but he is still subject to the limitations of his body. The body in itself is not problematic, because pleasure is a good thing. The problem is that the businessman has just *one* and one *aging, limited* body. Unlike the fisherman, he can escape limits of place and its bounded energy flows. But, like the fisherman, he cannot escape his body. With 3-D printed organs and genetic editing, though, we are pushing this frontier.

The logical conclusion of the story is not just decoupling and convenience. The logic goes further to what the transhumanists call the principle of morphological freedom: that one can take on any and every form. Indeed, this is written into the modern conception of energy as something that stays the same regardless of any changes in form. The 'self' that experiences progress takes on the same formlessness—as a shape-shifting essence unattached to any incarnation. It can be converted, like E, into this form or that. Locke's project of liberalism is at work here—to be liberated from the claims of tradition, place, family, class, and body.

This self will experience levels of pleasure inaccessible within the limits of our currently embodied form. Our bodies, after all, are the product of the same evolutionary process that created the energetically inefficient leaves that scientists are improving. Why would we think this blind process would maximize our experiences of pleasure? And why would we enhance plants but not ourselves? The ultimate power of the Fates is to cut the cloth of our lives short. Why wouldn't we fight death?

The transhumanist and US presidential candidate Zoltan Istvan (2019) sketches a Promethean 2.0 future. In the "nanotechnology era…we can recreate environments as we please, including those that are destroyed." Istvan imagines that nanotech and genetic editing techniques will allow us to "create whole mature forests and jungles in a week's time." The 'nature' we create won't be cruel like nature 1.0. Rather, we'll "create new environments that fit our modern needs. These will be virtual, synthetic, and machine worlds." The Eden we recreate will not be Earth-bound. We could cram all of humanity on a microchip that sips electrons to produce infinite and infinitely changeable and pleasing worlds. Now that's radical decarbonization—ditching the carbon body altogether for a silicon upgrade! We can also build worlds on other planets. If we can shape-shift out of our bodies, why not ditch the third rock?

We are entering the citadel of knowledge and gaining the ability to make any thought become a reality. The project of controlling fate is part of a cosmology that sees human intelligence (our divinely gifted fire) not as a short-lived quirk in an indifferent universe. Rather, we will come to dictate the fate of the universe. Given the law of accelerating returns, our knowledge is expanding exponentially. Remember, we are on the verge of mastering qubits. Leading transhumanist Ray Kurzweil (2018) argues that our "ultimate destiny" is to infuse the universe with our intelligence. We will do that by reorganizing matter and energy into 'computronium.' Rocks, trees, lakes, stars—everything will come alive with our divine fire. The universe will awaken into a consciousness trillions of times more intelligent than anything we are capable of yet. We will indeed fathom the world we are making, because the world will become a manifestation of our intelligence. We will master fate by effectively becoming One with the universe, by *becoming Fate*.

Recall the theological doctrine of univocity. It states that between the powers of humans and God there is only a difference of degree, not kind. We are made in the image and likeness of God and we are becoming God, as it were, on the installment plan—incrementally but also exponentially through technological acceleration. Another way of seeing this is in terms of the ultimate magic trick of God: *creatio ex nihilo*. If efficiency means more productive output for less input, or more from less, then the endpoint is *everything* from *nothing*. The ultimate Promethean transgression is to break, or at least infinitely bend, the first law of thermodynamics.

Bibliography

Brassier, Ray. 2014. Prometheanism and Its Critics. In *Accelerate: The Accelerationist Reader*, ed. Robin Mackay and Armen Avanessian, 467–488. Falmouth: Urbanomic.

de Beauvoir, Simon. 2004. *Philosophical Writings*, ed. Margaret A. Simons. Champaign: University of Illinois Press.

Istvan, Zoltan. 2019. Environmentalists Are Wrong: Nature Isn't Sacred and We Should Replace It. *Maven Round Table*, April 13. https://mavenroundtable.io/transhumanistwager/transhumanism/environmentalists-are-wrong-nature-isn-t-sacred-and-we-should-replace-it-TZ7Msb4mOk-B3n4kNqsyqg/

Kurzweil, Ray. 2018. What Will Happen After the Singularity? *YouTube*, April 22. https://www.youtube.com/watch?v=lAJkDrBCA6k

Mill, John Stuart. 1861 (2001). *Utilitarianism*, ed. George Sher. New York: Hackett Publishing Co.

Pinker, Steven. 2018. *Enlightenment Now: The Case for Reason, Science, Humanism, and Progress*. New York: Viking.

Plato. 1996. *Protagoras*. Oxford: Oxford University Press.

Strauss, Leo. 1952. *Persecution and the Art of Writing*. Glencoe: The Free Press.

Part III

Energy Heterodox

Look at her looking. Forever with that mirror. What does she see? The pink lips, the red cheeks, the wide eyes, studying the round face, the tiny pout. Seeing all the parts and none of the whole.
Stephen Sondheim (1984)

13

Look at the Beaver Looking

*When I was a child, I spake as a child, I understood as a child, I thought as
a child: but when I became a man, I put away childish things.*
1 Corinthians 13:11

Dear reader, can I confess that my faith is waffling? Can we whisper our
doubts? Do you share them? I see you are still straight-faced, sitting at
attention in the pew of the orthodox church. Don't worry, all this *security*
hasn't morphed into *surveillance*. Big Brother isn't watching you. Not yet,
or is he? OK, then don't turn your head. Just blink twice if you think
things are going crazy. Blink twice if you think this sober rationality of
the orthodoxy has come unhinged that our stone-cold machines are playing in the clowns. When did our church tent get so colorful, so whimsi-
cal? *Computronium?!* I know: We cannot abandon our faith, because it is
our way of life. "I believe because it is absurd," wrote Tertullian.

Still, though, blink twice. I'll understand.

If I can whisper some more, I will tell you when I started to lose faith
in the orthodoxy. It was on the drive down highway I-35, the clogged
aorta of the American heartland. I was taking my son Max to gymnastics

© The Author(s) 2021
A. Briggle, *Thinking Through Climate Change*, Palgrave Studies in the Future of
Humanity and its Successors, https://doi.org/10.1007/978-3-030-53587-2_13

with his little sister Lulu strapped into the booster seat beside him. It's a busy stretch of road through Denton, a city bursting at the seams with new homes, new apartments, and new shopping malls. The highway is being widened but there is still a bottleneck near the gym. Here is Locke's *increase* as experienced in daily life. Oh, the traffic jams! The road rage of all these beneficiaries of development.

To get Max to practice, I have to creep through the bottleneck and then fly downhill in the minivan only to slam on the brakes to make the sharp turn into the decrepit parking lot. To exit the gym, you have to get to warp speed quickly to match the flow of traffic. The drive is not safe. To attempt to walk or bicycle there would be suicide. The road is littered with 'snarge,' the bodies of raccoons, possums, and armadillos. Cars sitting still are fine. People are nice. But the hybrid car-person is a loud, prairie-paving, possum-killing, air-defiling impatient asshole.

Just before that final downhill, I see the glowing circular sign rising over the ridge like a yellow moon: the Buc-ee's beaver. If you are not from Texas, you may not understand Buc-ee's—this chain of *destination gas stations*. At the Denton location there are *ninety-six gas pumps* stretching nearly a quarter mile. It is enormous. But people don't just gas up. They get the mugs, the T-shirts, the beef jerky, the baby onesies, and the "beaver nuggets," a snack of puffed corn and corn syrup.

All of it is emblazoned with the mascot: A grinning buck-toothed, doe-eyed beaver wearing a red cap gazing vacuously upward. Like the Mona Lisa, how much mystery is packed into that look! What do you see up there, dear beaver? The next flavor of beef jerky? An even larger soda? Next fall's line-up of retro kitchen wall décor? A perpetual orgasm?

My faith wavers: is this the totem animal for the Anthropocene? A humanized beaver that chops down post-oak forests to build convenience stores. Ack, didn't we say that our faith was all about *convenience*? But are we sure we meant *this*? Are we still progressing or are we slipping back into juvenilia? Is high energy infantilizing us? The transhumanists tell us that our species is maturing, but why does it so often look like we are just extending childhood? Cruises, theme parks, discount vacation packages, streaming entertainment, and so on. When do we put away our childish things? What would we pick up in their stead? If it is heavy, would we still be strong enough to hold it?

The Buc-ee's billboard reads:

SOME SAY BEST JERKY EVER

But does anyone say BEST CULTURE EVER? I know: focus on factor M. Certainly no other culture in history has had the Earth-moving machines that this beaver was able to command. No other culture can raise and widen the highway to allow for this constant stream of semi-trucks heading to Walmart, Target, and Kroger. Truckloads of beaver nuggets. Is that not a *sure sign of progress*, as glaring as the billboard? Remember E X T = P. Energy multiplied by technology gives us what we want: products. No culture has ever had more P!

What are these, tears?! Am I laughing or crying? Now I must whisper more softly. Could it be that our faith made perfect sense in the eighteenth century but now its excesses and contradictions have started getting the best of it? Yes to more and more technology and yes to capitalism and yes to individualism when Hume was staring at a landscape of rocks and sheep and sheepherders who died young. But now staring into the languid eyes of the Beaver of Progress, should we say 'maybe' or should we request a new Council of Elders to reconsider church dogma in light of the world that the church has made?

But our Elders, our elites of techno-capitalism, are too busy folding their golden parachutes and stocking their bunkers for the coming apocalypse. They preach innovation and risk-taking, but they are playing it safe. They have an escape plan (Latour 2018). They can afford it. The rest of us, though? I guess eventually we'll all be billionaires with our own bunkers. Until then, relax and enjoy the beaver nuggets.

From the orthodox scripture: the solution to the problems caused by modernity is more modernity. But what is solution and what is problem? That beaver is so popular *and* so deplorable. What we call 'development' we might also call 'neoteny' or the reversal of development otherwise known as juvenilization. Decadence is the paradox of a declining culture on an upward technological trajectory.

I need to get a hold of myself. I've been thinking some heterodox thoughts. Just because everyone in the "energy sector" is so serious doesn't prove that the emperor is in fact wearing clothes. That story shows us how the most outlandish things demand the most earnest devotion! We're talking about dumping sulfates into the stratosphere to mitigate climate change; where is the child who will point a finger and laugh and laugh?

Then again, the heterodox thinkers in the following chapters might be the real clowns. There are many names for the clown: a misanthrope, a Luddite, or an eco-fascist who wants to ration bacon and air travel. A hypocrite, a curmudgeon, a philosopher, or a fool who romanticizes premodern ways of life, which everyone knows are nasty, poor, brutish, and short. I need to start over, again!

*   *   *

Let's revisit the parable of the fisherman, that icon of 'under-development.' The moral of the story was that in lacking energy, he lacked the security needed to truly enjoy life. Modern energy services (a powerboat, the electricity grid, etc.) would help him to control fate, liberate him from burdens, and enrich him with wealth and commodities.

Anthropological studies of hunter-gatherer cultures, however, paint a more complex picture (Sahlins 1972). Hunter-gatherer lives often compare favorably in terms of leisure with lives led by people in affluent, industrialized societies. It's true that hunter-gatherers would not have the advantages of modern medicine, but neither do they have the disadvantages such as antibiotic resistant super-bacteria or prolonged periods of decline and dying in impersonal hospital settings. Hunter-gatherers don't have in-home entertainment systems. But over 80% of people in modern society have jobs that bring them no joy in order to earn the money to pay for their Netflix binge until their next slog at work. Hunter-gatherers don't have well-lit streets. But they do have the starry skies we pay big bucks to enjoy on rare occasions.

The opening parable of the orthodoxy portrayed a 'conversation' between the businessman and the fisherman. But in real life the encounters of modernity have been more about violence (see Dunbar-Ortiz 2014). The 'businessman' plunders and dispossesses the 'fisherman,' creating the modernized poverty that modernity is then deployed to solve. Most subsistence cultures, in fact, did not have low EROIs (energy returned on investment) as assumed by the parable. Indeed, it is the

EROI of our so-called advanced extractive culture that is steadily declining. Andrew Nikiforuk (2012) notes that in 1919 it took one barrel of oil to net 1200 barrels. Now that ratio is down closer to 10 or 5 to 1, which puts it lower than many subsistence cultures. We are the ones working harder and harder to stay in the same place.

This is not to deny the existence of "alternative modernities" around the world complete with fossil fuel infrastructures. Burma had something of an oil industry in place a hundred years before 'Colonel' Drake struck it rich in Pennsylvania. The point is that, as Amitav Ghosh writes, imperial rule is what assured that the many variants of development going on around the world "came to be suppressed, incorporated, and appropriated into what is now a single, dominant model" (2017, p. 108). The imperative of growth, in other words, seems less like a universal law governing all civilizations and more like the obsession of just one. That energy growth has taken a global hold, then, tells us that this isn't just a story about some universal human potential coming to fruition in a process of 'development.' Rather, modern energy is a story about the vectors of cultural dominance: imperialism, slavery, colonialism, and capitalism. It is what Serge Latouche (1996) calls "the westernization of the world."

The problem with the orthodoxy is rooted in White's factor-M-for-means and factor-E-for-ends dichotomy. This is a distinctly *modern* dualism that he attributes to *all* cultures. The modern energy paradigm dis-embeds practices from their contexts and separates them into 'means' and 'ends.' Are we to call the hunt for bison a mere 'means' for survival? Would it be unequivocally better if they could just grab a pound of ground beef at the local Walmart or zap some buffalo nuggets in the microwave? It's more convenient, after all! Are the hunters just clocking in for another day of 'work'? No, the hunt is a ritual-rite-means-end tapestry that scrambles any simple dualism.

Even for us there is leakage across these categories. As if practices, say, for gathering or growing food are not simultaneously ends and means. Preparing a homemade meal is not merely an instrumental act—one that machines should render faster, more efficient, and more convenient. Though it is not counted by any metrics used by the Department of Energy or the Energy Information Administration, we have to admit that a rushed meal consisting of hot dogs and chips from the convenience store does not *add up* to much good. A meal in a pill would be most efficient. But do we only want to save time or also *savor*?

From the modern energy paradigm, agriculture is measured by units of grain per hectare. Transport is measured by miles per hour. Lost from this view are the enormous qualitative changes that result from scaling up the size of farms or boosting the speed of vehicles. *More* is our way of life, but more isn't just more; more is different.

There is no adequate way to count the shift from the old town square or Main Street to the strip mall on the highway. There is no simple metric for the loss of character in the resulting "geography of nowhere" (Kunstler 1993). Yet you can *feel* the foreboding and dehumanizing landscape of mega-gas-stations. Isn't that feeling the *real deal*? What about the feelings of the dairy farmer in Wisconsin holding back tears on the radio as he explains being forced to sell his cows due to globalization and automation? "I love these fricking cows," he says, "more than anything else. I don't want to do anything else, you know?" (Simon 2019). The energy orthodoxy gives only a blank stare, as if to say, "You are free now from the land, from those pesky animals. Smile like the Beaver of Progress."

If it cannot be counted, it is not real, right? That explains how the orthodoxy can swallow it all with a straight face and pronounce it all to be *progress*. Yes a six-year-old child is making millions on YouTube by reviewing toys. Sure grown men have millions of followers for playing video games or even just for commenting on other people playing video games. At touchofmodern.com you can buy a levitating football helmet lamp to show off your team pride. So what? The numbers are going up. Wealth is increasing. You can count that. But, can you see why I am flinching? Isn't this narcissism, nihilism, decadence, malaise, an untamed id, childishness?

Bill Gates is that philanthropic businessman from our parable bringing energy to those still suffering under the starry skies. Remember his formula for decarbonizing the economy while continuing to grow it: P x S x E x C = CO_2. 'S' is for services, which "should go up." It "is everything: food, clothing, heat, houses, cars, TV, toothbrushes, Elmo dolls, Taylor Swift albums, etc."

There's that problematic 'etc.' that et cetera of "to infinity and beyond." The orthodoxy operates at such a high level of abstraction that Elmo dolls and TV are treated exactly the same as food and heat. All qualitative distinctions have been eliminated. It is a form of brute utilitarianism worthy of Jeremy Bentham who wrote: "Quantity of pleasure being equal, pushpin is as good as poetry." All values, all commodities, are equal. All that matters is: more.

For Gates, clothing is treated the same whether it is the first pair of shoes or the fortieth. Here is the key blind spot of the orthodoxy: it has no sense of *thresholds*. Its logic is purely linear. Heterodox perspectives show the non-linear shapes of our high-energy lives: the shapes that *flatten out* with diminishing returns. For example, Robert Gorden (2016) argues that the age of rapid technological innovation and economic growth is over. The major, life-transforming inventions are done and we are left with increasingly marginal tweaks. It's another way to talk about a diminishing EROI—pumping more and more energy into innovations that do less and less to elevate the human condition.

The heterodox truths of our age also reveal shapes that don't just flatten out—they *reverse course* as more goes from being better to being worse. The orthodox faith in *productivity* needs to be balanced with a heterodox awareness of *counter-productivity*. Henry David Thoreau is a good touchstone for this point. The fisherman practices the philosophical life of voluntary poverty that Thoreau preaches. We are constantly solving problems, he says, with formulas more complicated than the problems themselves: "With consummate skill he has set his trap with a hair spring to catch comfort and independence, and then, as he turned away, got his own leg into it." Thoreau spends a lot of time discussing the 'economy' of food, clothing, and shelter, because he wants us to see how easily we can slip from mastering these necessities to being mastered by them.

We must be attentive to those inflection points or thresholds—the golden mean in the canon of virtues—where we have *enough*. The tricky part is spotting when we have slipped into excess. Without this discernment, our *needs* just keep growing and growing and we get tangled in the 'superfluities' that have somehow become necessary. The speaking creature now needs fiber optic cables under the ocean and satellites in space to communicate. The businessman isn't the wise one. He is falling down an infinite regress. He is trapped on a treadmill going faster and faster and getting nowhere. He offers a recipe for endless distractions and a life of forgetfulness, that is, of losing sight of ultimate meaning and purpose. In short, it is nihilism—a life of "quiet desperation."

* * *

The Stephen Sondheim quote at the beginning of this section is from a musical inspired by a George Seurat painting. Seurat was a pointillist who used small distinct dots of color in patterns to form images. This creates a superposition like the duck-rabbit image where what one sees depends on how one is looking. To zoom in on the dots is to see "all the parts and none of the whole." This is the reductionism of the energy orthodoxy, which multiplies and counts the dots yet ignores the emerging images. In itself, this is not problematic. The problem, and the absurdity, is the totalizing narrative to accompany this point of view—the insistence that this is *the only legitimate point of view*. The belief that *there is no emergent pattern*. That more dots—more P, more S, more energy, more stuff—only adds up to more freedom and more choice and thus is *always* better.

Modern science is caught up in a wicked superposition here. For the orthodoxy it is the winning formula of Enlightenment. I won't deny the truth of this. I am not going to turn into a flat-earther! My children are vaccinated. I love Bill Nye and "Cosmos" (with Sagan and Neil deGrasse Tyson) just as much as anyone. I am sympathetic with the March for Science movement. But which science? Chemical weapons, facial recognition software, "clean coal"? Where are we marching?

The *other* truth—the heterodoxy—about modern science is reductionism or 'scientism' when human phenomena are treated as wholly explicable in terms established by the physical sciences (see Olson 2008). Even if the intent is to find a unifying 'consilience' of different kinds of knowledge (e.g., Wilson 1998) or a perfect symmetry, it often ends up with a portrayal of lived experience that is as empty as the eyes of the Beaver of Progress. Love becomes hormones. Communities become networks. Justice becomes algorithms. Energy becomes a service and a commodity. Mother Earth becomes a heat engine, a matrix of forces, something much harder to care for and love. The aesthetics is all bleached out. If we don't attend to this other truth, we'll end up on perpetual space cruises like the denizens of *Wall-E*, idly slurping smoothies while playing Candy Crush.

This is what William James decried in his speech on "The Energies of Men." Later, on his death bed, James criticized the scientism of Henry Adams' 1910 essay "A Letter to American Teachers of History." Impressed with the new science of E, Adams ran with the impulse of scientism. Thinking can be explained in terms of electricity and chemicals. History must follow the second law of thermodynamics. People are "human molecules." The humanities must bend to fit the reality disclosed by the sciences. Those ineffable "vital energies" behind love, hate, politics, courage, and play are banished to unknowable subjectivity.

As the philosopher Albert Borgmann (1984) notes, modern science is characterized by an *apodeictic* mode of explanation where the same covering law explains all instances. This is a move from diversity to sameness. It shows us that beneath the apparent differences between bread and bronze there is the same atomic matrix following the same universal laws of motion. This is that "perfect similarity" discussed before. It is the E of a universal currency. Obviously, this gives science a coherent and detailed view of reality and enormous power to cut and paste its pieces.

Yet the scientific mode of explanation is also a step away from human intelligibility. By erasing the uniqueness of bread and bronze it is poorly equipped to handle the aesthetic work of *world articulation*, that is, of pointing out significance and meaning. This is what I gather from William Blake when he wrote: "May God us keep/ From Single vision and Newton's sleep." Science alone offers no criterion to know when to stop on the chain of being from universe through galaxy through solar system through Earth through bread and bronze through molecules through atoms through quarks. Science offers no account of relevance: it cannot tell us which problems are worthy of attention. It can help us to transform nature, but cannot tell us how much or in which ways to transform it.

It's no wonder there is no limit on the energy orthodoxy! It's based on modern science, which like the Beaver of Progress, is always staring over the horizon. Modern science (unlike Aristotle's science) gives no answer about when to stop. So too, the energy orthodoxy has no stopper. Even when we have obviously drifted into decadence, excess, and childish hyper-consumption, we keep inventing new markets and new gadgets. Can your pants make a phone call? Can your shirt communicate with your fridge? You are so primitive! The vital energy of desire—that we refuse to reason publicly about—becomes the elephant in the room. It becomes the prancing, naked emperor engaged in ever-more foolish displays that everyone calmly accepts as completely normal and indeed necessary!

The spreadsheets at the Department of Energy or the speakers at the CERAWeek conference on energy hardly even mention the ends of energy. There is no articulation of that world that we actually inhabit— that world of minivans and grinning beavers. All of that is washed out in the numbers and graphs. Lots of graphs could show you numbers and lines trending upward with the building of Buc-ee's. Economic growth, energy consumption, consumer choice, tax rolls, and so on. But the numbers stand mute at the gas pump in the sweltering heat with the din of traffic echoing off of the concrete. They just give a shrug. If they could speak, all they would say is: "I dunno! What do you think?"

At least White mentioned factor-E-for-ends. He nods at the human lifeworld. But to speak of song, dance, companionship, 'etc.' (there is that et cetera again) is to paint culture with the broad strokes of a modern abstract artist. More precisely, it is to measure culture with the brute implement of GDP (see Stiglitz 2019). It all comes down to P! This is in keeping with an underlying account of what is properly scientific and real. What is dance, after all, but the movement of limbs? Song is the creation of sound waves. Companionship is matter moving in closer proximity to other matter. Recall from White: a men's club is a men's club, whether in the culture of the Zuni, the Han, or Victorian England. It's all the same! After all, F *always* equals ma.

We have to come down from the clouds of numeracy and abstractions to get to the plane of lived experience where questions about meaning and flourishing become live issues for us. This is the heterodox move. In so doing, we see what is not controlled by the orthodoxy and counted by its metrics. As a friend and translator of the *Tao Te Ching* Lao Tzu puts it, the world of E with its numbers and graphs is the world of the *just-so*, where things appear to be fully explained and accounted for (Brown 2019). There is also, however, the world of the *more-so*, which abides in and through us like the Force but can only be gestured at and not fully controlled or measured. The first poem reads:

> The tao that can be told
> is not the eternal Tao.
> The way that can be put into words
> is not the Way.

To speak of "the Way" is bring us back to the Fates. In Anglo-Saxon cultures, the word for fate was *wyrd* (as in the witches or Weird Sisters in *Macbeth*), which has since transmogrified into our word 'weird.' In Old English *wyrd* meant "to come to pass, to become," which carries echoes of older energetic terms related to potentiality and becoming. *Wyrd* in Germanic tongues is related to *werden* or *worden* (to be). The *wyrd* like the Way is "that which is in the process of happening." Of course we can control fate, but there are laws of equal and opposite reactions—each act of control brings an act of *wyrding*.

In western philosophy, to pay attention to "the process of happening" is known as phenomenology or the study of life-as-it-is-experienced. This is where we can find energy heterodoxies, in the world as it appears to us unmediated by metrics. Enough with *counting* things—all this P for productivity—let us turn "to the things themselves." We'll begin, naturally, with a pooping unicorn.

Bibliography

Adams, Henry. 1910. *A Letter to American Teachers of History*. Washington: Press of J.H. Furst Co.

Borgmann, Albert. 1984. *Technology and the Character of Contemporary Life*. Chicago: University of Chicago Press.

Brown, Keith Maggie. 2019. "The Heirophant: There is More-so than the Just-so." Call Me Maggie. Blog. https://callmemaggie.com/2019/06/04/hierophant-more-so-than-just-so/

Dunbar-Ortiz, Roxanne. 2014. *An Indigenous Peoples' History of the United States*. Boston: Beacon Press.

Ghosh, Amitav. 2017. *The Great Derangement: Climate Change and the Unthinkable*. Chicago: University of Chicago Press.

Gordon, Robert. 2016. *The Rise and Fall of American Growth: The US Standard of Living since the Civil War*. Princeton: Princeton University Press.

Kunstler, James. 1993. *The Geography of Nowhere: The Rise and Decline of America's Man-Made Landscape*. New York: Touchstone.

Latouche, Serge. 1996. *The Westernization of the World: Significance, Scope, and Limits of the Drive Towards Global Uniformity*. Malden: Polity Press.

Latour, Bruno. 2018. *Down to Earth: Politics in the New Climatic Regime*. Medford: Polity Press.

Nikiforuk, Andrew. 2012. *The Energy of Slaves: Oil and the New Servitude*. Vancouver: Greystone Books.

Olson, Richard. 2008. *Science and Scientism in Nineteenth-Century Europe*. Champaign: University of Illinois Press.

Sahlins, Marshall. 1972. *Stone Age Economics*. Chicago: Aldine Publishing.

Simon, Scott. 2019. Physically and Mentally Draining: The Economic Hardships US Dairy Farmers Face. *National Public Radio*, April 20. https://www.npr.org/2019/04/20/715393992/physically-and-mentally-draining-the-economic-hardships-u-s-dairy-farmers-face

Sondheim, Stephen. 1984. *Sunday in the Park with George*. Broadway: Musical.

Stiglitz, Joseph. 2019. It's Time to Retire Metrics like GDP. They Don't Measure Everything that Matters. *The Guardian*, November 24. https://www.theguardian.com/commentisfree/2019/nov/24/metrics-gdp-economic-performance-social-progress

Wilson, E.O. 1998. *Consilience: The Unity of Knowledge*. New York: Vintage Books.

14

Invention Is the Mother of Necessity

[N]eeds are much more cruel than tyrants
Ivan Illich (*1992*)

Necessity is the Fates. It is the limits imposed by a particular body, place, or season. The orthodoxy believes that necessity is the mother of invention. Prometheus stole fire from the gods, because we were naked and shoeless. We use this divine energetic spark to rebel against necessity and win our freedom. It's another half-truth. The heterodox perspective is also true: invention is the mother of necessity. Or, better said, it is the mother of *needs*. Where the orthodoxy rules, we may struggle less with the Fates and their fixed limits, but we suffer more from the Machine and the needs it keeps piling on top of us.

My daughter Lulu's kindergarten class once did an exercise on "needs and wants." You might think it would be a black-and-white exercise. Food and shelter are needs. Candy and mansions are wants. "But," the wily kindergarteners respond, "isn't candy a kind of food? And a mansion is just a large shelter, no?"

© The Author(s) 2021
A. Briggle, *Thinking Through Climate Change*, Palgrave Studies in the Future of Humanity and its Successors, https://doi.org/10.1007/978-3-030-53587-2_14

"Ok," the exasperated teacher could respond. "Fine, but take this toy!" She gestures at Poopsie the pooping unicorn. A child brought the toy unicorn to class as an example of a 'want.' Poopsie is a plastic doll that retails at Target for $49.99. You feed her with her custom spoon and bottle and she 'poops' out a shimmery slime that you can play with. You can even use single-use foil packets of "unicorn food" to add extra sparkles and glitter to the unicorn excrement. The teacher concludes: "This is a want. No one *needs* a pooping unicorn!"

The class is compelled to agree. But then the economist enters and introduces the paradox of thrift. "Yes, kids, you *could* elect to save money and not buy the unicorn, but did you think about the workers who make Poopsie? They *need* their jobs to buy food." What about the workers who make the boxes for Poopsie, the truck drivers who bring Poopsie to the store, and the employee at the cash register? Everyone who participates in the *convening* of Poopsie *needs* people to buy the pooping unicorn and all the other P-for-products and S-for-services that we invent.

A society premised on Locke's doctrine of *increase* (economic growth) operates by the imperative of growing needs. What was a luxury yesterday is a need today. Computers were nearly inconceivable a hundred years ago, but now our way of life depends on them—so we *need* to keep the energy flowing. My minivan would have been a wonder to the pioneering Rayzor family that took a covered wagon to Denton in 1866. I don't wonder at it; I need it to get the groceries. Though Sam Houston and Jim Bowie won Texas independence without oil, modern Texans need 2-ton trucks to experience their freedoms.

Obesity from beaver nuggets, kids demanding pooping unicorns that people in nice suits spent hours of serious labor designing, and the inability to get around town without a car loan. These may be "first world problems" but the 'problems' are neither superficial nor accidental. There is a logic holding them together: They are what happens once we understand energy as something we *lack,* that is, as a "basic need" that must be provided to us. The logic of lack is endless—the hole can never be filled.

As needs grow, progress starts to look like the elaboration of ever-more intense forms of *poverty*—we never have enough! At a Heartland Institute conference on energy, I attended a panel titled "Energy and Prosperity" (Heartland Institute 2017). Of course, prosperity can mean many things.

(Recall the question of what the elephant *is* and how to measure it.) A major study, for example, found that prosperity comes from community ties and meaningful relationships (Waldinger 2015). Prosperity can also mean skillful engagement with the world, access to fine arts, a pleasing neighborhood, time spent devoted to noble causes, or spiritual peace. True orthodox believers, though, the panelists set aside the question of what to *count as* prosperity to get right to the counting. They naturally talked about how GDP and energy track together linearly upward. More E-for-energy means more P-for-prosperity.

But they missed their own punchline. The US economy is twenty times larger today than it was a hundred years ago. Yet by some measures, 30% of Americans still suffer from energy poverty (Ingber 2018). How can you have a cell phone—a computer—in your pocket but still be poor? Because you *need* it and so much more!

Happiness, too, fails to follow the imperative of linear growth in P. This has long been the message of the energy sector's resident heterodox thinker, Vaclav Smil. Distinguished Professor Emeritus at the University of Manitoba and Fellow of the Royal Society of Canada, Smil has authored some forty books about energy. The main source of Bill Gates' impressive knowledge about energy is Smil. Gates has confessed that he waits for new Smil books the way others might wait for the next *Star Wars* movie.

Yet Gates has not absorbed the central teaching of his master: we need to put a stop to the logic of *increase*. This is not just because we are bound to hit an ecological wall. It's also because *after a certain point* more energy fails to deliver more well-being. Smil writes, "We could all be perfectly happy living at the level of consumption and income as Frenchmen in 1959" (in Voosen 2018). Elsewhere, he concludes that "the quest for ever higher energy throughputs has entered a decidedly counterproductive stage, insofar as further increases of per capita energy use are not associated with any important gains in physical quality of life or with greater security, probity, freedom, or happiness" (Smil 2010, p. 726). Figure 14.1 shows in graph form what Smil and others have found in empirical studies about the diminishing returns of energy consumption.

There is a saturation point to human flourishing just as Aristotle imagined with his doctrine of the mean. There can be too little E but also *too*

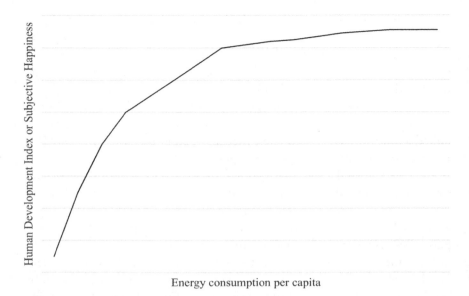

Fig. 14.1 The diminishing returns of energy consumption

much. Smil wryly notes that the only guaranteed outcome of greater energy inputs is environmental ruin. We are altering the climate for no good reason other than the logic of E, which starts out beneficent but then twists into something comical and tragic—those needy people at the altar of grinning beavers, pooping unicorns, and endless desires for tourism.

Smil calculates that Americans and Canadians consume 8 TOEs (tonnes oil equivalent) of energy per person per year. Yet "no indicator of high quality of life – very low infant mortality, long average life expectancy, plentiful food, good housing, or ready access to all levels of education – shows any substantial gain" once the average energy consumption rises above 2.5 TOEs per person per year (Smil 2010, p. 721). That's an incredible overshoot. Smil draws the ultimate heterodox conclusion that it is time for "determined moves to end the historic quest for ever higher energy throughputs, to put in place rational limits" (p. 728). Limits?! Is he nuts?

Then again, Americans command the energy equivalent of ninety-three slaves. That's another way of saying that we *need* a lot of machines. As the saying goes "as many enemies as you have slaves." Maybe we've crossed a threshold where things shift and freedom comes not from transgressing limits but embracing them.

* * *

My grandmother Mary was born in 1913. When she was a child, less than half the homes in the United States had electricity, and air conditioning was almost unheard of. She was born into a world of animal energies—America's herd of draft animals peaked at 26.7 million in 1918. Gasoline-powered tractors and other forms of fossil-fueled mechanization were just starting to take over. The agricultural labor force was declining, but 30% of Americans still worked on the farm. Mary had a limited wardrobe and clothes were washed infrequently.

My daughter Lulu was born about a hundred years after Mary. Consider the magnitude of change in just four generations. Electricity is ubiquitous. The same goes for air conditioning. The agricultural labor force has shrunk down to 2% of Americans. The world of animal energies is gone. Human population is growing by a billion every twelve years. Americans spent twice as much on consumer goods in 2017 than they did just fifteen years earlier. With reams of personal data and the 5G internet, Amazon is perfecting the art of seduction. The average American throws away 81 pounds of clothing every year (Gilmore 2018).

Every minute, over three hundred hours of videos are loaded onto YouTube. Lulu navigates both Walmart and the online world with a yawn. When she plays games on my phone, she quickly tires of them and wants to download another one from the infinite apps available on the play store. Mary would have been agog just walking the aisles of Walmart and looking at the mega-packs of toilet paper made from the virgin pulp of Canadian pine trees. I wonder if she could even make sense of the endless scroll of the Facebook newsfeed.

Many energy conversions on Earth are transformations of sunlight. Wind and water flows are nearly immediate. Food is stored solar energy over a few weeks or months. Draft animals are a few years' worth of stored solar. A mature tree is a few decades of sun. Fossil fuels store solar energy from millions of years ago. The irony here, in other words, is that by tapping into deep time, we have erased our sense of history. We forget to be grateful for the bounty, because it doesn't *appear to us as bounty.* Aluminum was once so valuable that Napoleon III reserved his aluminum dinnerware for occasions of state. Electricity made refining aluminum cheap. Now we wrap our leftover pizza in aluminum foil. It is convenient, but maybe we are forgetting how to treat the world *as precious.*

Things that are ordinary today were inconceivable yesterday, and tomorrow will bring something else presently unknowable. For the orthodoxy, this is progress, which is an understandable view. I don't want to give up my way of life. As I type this, I am drinking Florida orange juice and eating a banana from Guatemala at my parent's house on a mountain in Colorado having driven here from Texas with Lulu and Max in just one day in our minivan. One of those March blizzards is coming, and I look forward to watching the snow from the warmth of a home built and supplied by vast energy networks.

I like this way of life. But Mary liked her way of life too. So did Laura Ingalls Wilder, author of *The Little House on the Prairie* set in the nineteenth century. And so did the Native Americans who fought against Laura's father and the other settlers and the benefits they promised from a 'civilized' existence. Progress for the orthodox world of the *just-so* entails movement closer to some goal posts. It can be measured as growing P from a growing factor M. A cultural relativist would deny that there can be any measurement of progress. A heterodox view tries to avoid both extremes to find a mean that is hard to define, because it abides in the *more-so.*

Mary was born in the early stages of the advertising industry. When capitalism started producing an overabundance of goods the imperative of *increase* required convincing people of their inadequacies (see Packard 1957; Ewen 1976). They lacked and were in need of so much; it's amazing they didn't notice it before! Kids in Lulu's generation are exposed to 5000 ads daily. As a result, Lulu finds herself drowning in desires and I find myself unable to park our minivan in the garage due to all the stuff we were convinced we needed. We have plastic tubs full of the riff-raff of a high-energy life of convenience. We've got old toy drones in those tubs that would have blown the minds of the pioneers. But those were *yesterday's* toys! They already fell through the abyss of desire. What's next?

The orthodoxy wants to separate the 'real' energy of E from the squishy energies of human consciousness. But all that E dug up in coal mines and stored in batteries from Lithium mines is driven by this growing pulse of human wants and needs. We can, then, turn to psychology for some of our most important energy concepts.

For example, consider "shifting baseline syndrome." In the absence of experience with historical conditions, members of each new generation accept the world they were raised in as normal. This applies to our present condition of material abundance and pooping unicorns. It also applies to conditions of environmental impoverishment (Soga and Gaston 2018). Many insect populations are crashing, but new generations adjust to forests devoid of lightning bugs or butterflies. What would have seemed pauperized to their parents seems comfortably familiar to them. Rachel Carson's silent spring would just be a regular spring for the next generation.

Another term is the "hedonic treadmill." People tend to return to a stable level of happiness after major positive or negative changes in their lives. We grow accustomed to whatever happens, or as Raskolnikov puts it: "Man can get used to anything, the beast!" The children in the *Little House in the Big Woods* received one handmade gift at Christmas and were thrilled with it. Lulu and Max each received thirty gifts last Christmas and they were not thirty times happier than Laura Ingalls Wilder with her new doll. In fact, they seemed rather unimpressed. All that love convenes in grey tubs in the garage and eventually the city dump.

The treadmill gives the illusion of progress, but is actually the investment of ever-greater amounts of energy to stay in the same place. Yet we can't give in to relativism. Some inventions do move us forward. Even Ed Abbey, the defender of wilderness, put refrigeration in that category. Other things, though, only represent an accelerated economic churn without any genuine contribution to human flourishing (see Sarewitz 1996). I would be miserable if I were transported back to the 1870s. But that doesn't mean Laura Ingalls Wilder sat around pining for microwaves, minivans, and Mickey Mouse. If that were the case, then you would be miserable because you don't have i-FLIZ. I don't know what i-FLIZ is, but they will have it in the future. They'll get used to it; in fact they will come to *need* it. And they will believe we led impoverished lives due to its absence.

In the history of lighting, there is a curious turn around 1800 when James Watt and Humphry Davy start playing with nitrous oxide (Rhodes 2018). This is part of the quest to find luminous gasses that could replace the depleting stock of whale oil. Davy, who got addicted, suggested that nitrous oxide could be an anesthetic. At the time, though, pain was often seen as "the voice of nature," a stimulant that kept people alive. Surgery involved not just the virtues of the surgeon but also those of the patient, in particular, bravery. One report of a mastectomy in 1811 noted that it was twenty minutes of uninterrupted screaming.

We now take anesthesia for the elimination of physical pain for granted. The next step is to abolish psychological pain. The transhumanist David Pearce (1995) says that we can achieve permanent bliss through neuro-stimulation, drugs, and genetic technology. Like the orthodoxy, he thinks of happiness as a subjective state of mind. He defends the logical conclusion of *convenience*, which is to directly *convene* pleasurable experiences in our minds. We are destined to be brains in vats experiencing levels of pleasure we cannot even imagine now. This is how to break the saturation point that keeps the hedonic treadmill spinning in the same spot. We must use E, the capacity to do work, to boost our capacity to feel pleasure!

If that doesn't sound like an ideal trajectory, then we need to draw a line somewhere along the logic of progress. We have to mark a threshold where *more* arcs in a superposition to become *less*. But where can a line be drawn between the shoeless ape and the brain in the vat? At beaver nuggets? Pooping unicorns? Virtual reality headsets? We could never have enough—or even too much—pleasure or convenience, right? I mean, that's the whole point of breaking the rule of the Fates with high energy. But who knew we would *need* so much stuff to make us happy?

*　*　*

There was no Poopsie when my grandmother Mary was a child. Having no pooping unicorn to choose, she didn't have that desire. Was she deprived or was she liberated? Mary also didn't have air conditioning as a child. Now ask the same question. When is there excess and when is there deficiency? The orthodoxy has no way to address that question. It's all light, no shadow.

Jean Jacques Rousseau was one of the first heterodox voices to see the shadows cast by the Enlightenment. In his 1750 *Discourse on the Sciences and Arts*, he argued that material gains undermine moral virtues. So-called progress causes a "softness of character" and transform us into "happy slaves" content to trade a richer sense of freedom for the comforts doled out by technocratic systems of control: "Need has raised thrones; the sciences and the arts have strengthened them."

Rousseau talked about the mimetic nature of desire as *amour propre*. As society gets more complex, people want to be ranked favorably *in comparison* with others around them. Happiness doesn't rise with affluence, because someone else always has more than you do. Virtue doesn't improve, because keeping up with the Joneses requires inauthentic misrepresentations of one's true needs and genuine self. You keep changing your desires and needs in step with the Joneses, who are running on the treadmill next to you. Thorstein Veblen (1899) would later call this "invidious consumption" and "conspicuous consumption." The Marxist social critic Herbert Marcuse (1964) said that advanced industrial society created "one dimensional man," a creature defined by external "false needs" and devoid of internal or authentic vital energies.

The best analysis of energy and needs comes from Ivan Illich, the heterodox Catholic priest. In his 1974 *Energy and Equity*, he uses automobiles to illustrate the non-linear logic of thresholds. At first, faster vehicles shorten distances and lighten burdens. This initial stage of genuine progress is what he elsewhere calls "the first watershed." But around 15 mph we reach a threshold, a "second watershed," where increasing speed starts to become *counter-productive*. The transportation industry begins to dictate the configuration of social space in ways that actually increase distances and burdens. Environmental pollution is matched with a kind of psychic pollution of frustration, impotence, and disutility. Time isn't saved, it's made scarcer: the average American spends forty-two hours annually sitting in traffic jams (Anderson 2016).

The car-centered development serviced by the Beaver of Progress puts neighborhoods far from grocery stores, doctors, and recreation. Once roads bring trucks to villages, local shops are erased by Walmart (surrounded by an acre of parking spaces and *one* bike rack). In our daily lives, Illich argues that we become "a new kind of waif…constantly absent from a destination [we] cannot reach on [our] own but must attain within the day." When you factor in daily commutes, the time required to pay for a car, insure it, maintain it, and fuel it, Smil (updating Illich's calculation) puts the average speed of cars at about 5 km/h, which is no faster than walking but a lot more expensive. Of course, people can't just walk to work, because they live in worlds where they *need* their cars. We suffer from *over-industrialization*.

Illich draws a distinction between *transit* ("movements that put human metabolic energy to use") and *transport* ("that mode of movement which relies on other sources of energy"). Cars are not just bigger bicycles. They introduce qualitative changes that are full of implications not just for personal freedom but also justice. It is the poor who are forced to lose the most of their time as they wait at the bus stops of under-funded public transport systems. Others who can barely afford cars get less reliable ones prone to breaking down and swallowing whatever meager savings they can manage. In a survey, 40% of Americans said they could not afford a $400 emergency expense (Van Dam 2019). The security supposedly provided by the high-energy way of life is far from equally distributed. Exposure to growing needs becomes a new form of poverty and insecurity. The coronavirus pandemic, at least in the United States, made this fact painfully obvious in the way it disproportionately impacted minorities and the working poor.

Automobiles exercise a "radical monopoly" over traffic by rendering transit by walking or bicycling either impossible or unsafe. Many are free to choose a Chevy or a Ford, but they *are not free* to opt out of the transport system. Car commercials often show the absurdity of a culture that celebrates this narrow sense of freedom. In one, the car becomes a red balloon lifting a beautiful young woman up into mountain glaciers. Maybe that's what the Beaver of Progress sees on the horizon—that balloon girl of our imaginary freedoms. Of course, even if you get to the mountains of Yellowstone there are still traffic jams!

The French yellow vest movement that started in 2018 was in response to a new gasoline tax. The tax was designed to address climate change, which you might expect would be well-received in the country that housed the Paris Climate Agreement two years earlier. Instead, the tax led to riots in the name of economic justice. The wealthy who live in the heart of expensive cities can afford to do without cars. But those forced to the outskirts of town by the increased cost of living *need* their cars to get to work. French President Emmanuel Macron was cast as an elitist for worrying that climate change would bring about *the end of the world*, when the modernized poor of France are worried about getting to *the end of the month* for the next paycheck.

Illich (1992) argues that to be human has always meant "communal submission to the rule of necessity" in a particular place and time. This 'necessity' is the Fates. By contrast, 'needs' are a modern invention that has been cloaked in the guise of a natural fact. The orthodoxy begins with the "basic needs" symbolized by the fisherman. The poor *lack energy* and this need becomes the basis of a rights claim. They have the right to development, which is a promise to "break the rule of necessity." Progress means liberation from the necessity of living within narrow limits.

Yet in breaking the rule of necessity, a new rule of need was introduced. This rule is that of an endless lack that compels people to achieve ever-escalating minimum levels of consumption. The Fates give way to new masters, the corporate and bureaucratic social systems, that operationalize new needs. Necessity is a condition that must be dealt with. Needs are a lack or a bottomless hole that can never be filled. And because needs become rights, justice becomes chained to an ever-growing productivity of needs. Each new invention becomes a new need and, thus, a new rights claim. We all have a right now to triple bypass heart surgery when past generations would have just chalked that up to fate. To service these needs-rights claims of justice, we have to keep growing the economy and the energy services that drive it. In the process, we invent yet more needs and rights. We discover the hole was even deeper.

Poverty in the economies of subsistence ruled by necessity meant living within limits. Modernized poverty, by contrast, means running on the treadmill faster and faster only to just keep scraping by. The yellow vest protestors in France enjoy a material abundance that my grandmother's generation could hardly fathom. Yet their lives are still precarious. They are the modern poor who cannot live outside of the needs imposed by the wage-economy but who also cannot rise above its bottom rungs no matter how furiously they climb ever upward with more labor and consumption.

Illich concludes that equity cannot be achieved through the linear logic of giving more and more to everyone. The logic of "not enough" just fuels more inequalities. The richest 20% of humanity consume 70% of total energy, while the poorest 25% consume just 2.5% of total energy. Getting everyone up to the energy standards of North Americans would require adding *five times* the current global energy supply. Smil (2010) concludes that this is "utterly impossible." Justice comes not from dreaming of everyone floating on their own red balloons. It comes from seeing the "too much." We must limit the energy consumed and the speed traveled by the richest members of society.

Bibliography

Anderson, Tom. 2016. Commuters Waste a Full Week in Traffic Each Year. *CNBC*, August 9. https://www.cnbc.com/2016/08/09/commuters-waste-a-full-week-in-traffic-each-year.html

Ewen, Stuart. 1976. *Captains of Consciousness: Advertising and the Social Roots of the Consumer Culture*. New York: McGraw-Hill.

Gilmore, Nicholas. 2018. Ready-to-Waste: America's Clothing Crisis. *Saturday Evening Post*, January 16. https://www.saturdayeveningpost.com/2018/01/ready-waste-americas-clothing-crisis/

Heartland Institute. 2017. America First Energy Conference. November 9. http://americafirstenergy.org/videos/

Illich, Ivan. 1974. *Energy and Equity*. New York: Harper & Row.

———. 1992. Needs. In *The Development Dictionary: A Guide to Knowledge as Power*, ed. Wolfgang Sachs. London: Zed Books.

Ingber, Sasha. 2018. 31 Percent of US Households have Trouble Paying Energy Bills. *National Public Radio*, September 19. https://www.npr.org/2018/09/19/649633468/31-percent-of-u-s-households-have-trouble-paying-energy-bills

Marcuse, Herbert. 1964. *One-Dimensional Man*. Boston: Beacon Press.

Packard, Vance. 1957. *The Hidden Persuaders*. New York: Longmans, Green and Co. Ltd.

Pearce, David. 1995. *The Hedonistic Imperative*. http://happymutations.com/ebooks/david-pearce-the-hedonistic-imperative.pdf

Rhodes, Richard. 2018. *Energy: A Human History*. New York: Simon and Schuster.

Rousseau, Jean Jacques. 1750. *Discourse on the Sciences and the Arts*. https://www.stmarys-ca.edu/sites/default/files/attachments/files/arts.pdf

Sarewitz, Daniel. 1996. *Frontiers of Illusion: Science, Technology, and the Politics of Progress*. Philadelphia: Temple University Press.

Smil, Vaclav. 2010. Science, Energy, Ethics, and Civilization. In *Visions of Discovery: New Light on Physics, Cosmology, and Consciousness*, ed. R.Y. Chiao, M.L. Cohen, A.J. Leggett, and C.L. Harper Jr., 709–729. Cambridge, MA: Cambridge University Press.

Soga, Masashi, and Kevin Gaston. 2018. Shifting Baseline Syndrome: Causes, Consequences, and Implications. *Frontiers in Ecology and the Environment* 16 (4): 222–230.

Van Dam, Andrew. 2019. Are Americans Benefiting from the Strong Economy – Aside from the Rich? A Fed Report Raises Questions. *Washington Post*, May 23. https://www.washingtonpost.com/us-policy/2019/05/23/americans-arent-ready-weather-sustained-downturn-new-report-shows/

Veblen, Thorstein. 1899. *The Theory of the Leisure Class: An Economic Study of Institutions*. London: George Allen & Unwin Ltd.

Voosen, Paul. 2018. The Realist. *Science* 359 (6382): 1320–1324.

Waldinger, Robert. 2015. What Makes a Good Life? Lessons from the Longest Study on Happiness. *TedxBeaconStreet*, November. https://www.ted.com/talks/robert_waldinger_what_makes_a_good_life_lessons_from_the_longest_study_on_happiness?language=en

15

E, Neutrality, and Democracy

People and motors do not move through the same kind of space
Ivan Illich (*1983*)

A central half-truth of the orthodoxy is that energy is neutral. If this were the whole truth, then indeed we would only require instrumental reasoning. We could limit our ethics to questions about the means—for example, what kind of renewable energy program to support. And we could set aside contentious debates about the good life. You like to hike and he likes car racing? The orthodoxy says: "Fine! There's room for all lifestyles. Indeed, as we master greater flows of E (volts, or modern energy services) we expand the buffet of life choices available." According to the orthodoxy, energy is good and more energy is better. And part of what makes it good is its neutrality, because that guarantees expanded freedom of choice. It's another paradox that goes unremarked: energy is simultaneously good (thus value-laden) and neutral (thus value-free).

Earlier, I put this in terms of a pointillist painting and how the orthodoxy sees all the multiplying dots but not the emergent patterns. This is *one* way to look at high-energy society. Indeed, it is the easiest way and

© The Author(s) 2021
A. Briggle, *Thinking Through Climate Change*, Palgrave Studies in the Future of Humanity and its Successors, https://doi.org/10.1007/978-3-030-53587-2_15

the way that lends itself best to quantification. We can count new power plants and additional megawatts. But as with a pointillist painting, this is certainly not the *only* or most meaningful point of view. There is, in fact, a pattern to the high-energy life. This means that adding more E isn't neutral. Rather, it shapes society and the kinds of lives that it fosters in particular ways. It doesn't just expand choices, it also constrains them; it forces them to be choices made from *within* a particular pattern.

Below, I will flesh out this argument that modern energy services are non-neutral. But first, let's acknowledge the rational and moral force of the neutrality thesis.

The coffee maker on the kitchen counter works the same whether it is animated by electrons generated from coal, gas, nuclear, hydro, solar, or wind. Earlier, we talked about the "perfect similarity" of E as the universal currency. That you could have a radical change of the means (say from coal to wind) without changing the ends (the coffee), speaks to this perfect similarity. The frog leg of Volta's experiment kicks whether plugged into its native brain or the wall socket.

This also demonstrates the strength of the energy-as-neutral thesis. Because if you are able to swap something out without making any noticeable change in the end result, then you are clearly working with something that is *neutral* in a very pragmatic sense. Indeed, this kind of swapping-out phenomenon is essential to the grand energy transition to a carbon-free economy. Our way of life (the ends) can remain unchanged even when the energetic means are wholly substituted for other ones. It's reminiscent of the ship of Theseus, which is replaced board by board over time until every single piece of it is different and new, yet it remains the same old ship! It's as if we could pull the tablecloth of the carbon economy out from under our way of life *and* slip a new carbon-free tablecloth underneath it without spilling a single glass of wine.

Recall that energy-as-E (the "perfect similarity") came about in a rebellion against Aristotle's *energeia*. Given that *energeia* was woven into the fabric of Aristotle's ethical and political project, it shouldn't be surprising that the energy-as-neutral thesis has broader ramifications. Turning to those wider dimensions briefly will further strengthen the idea of neutrality as well as demonstrate its importance.

Aristotle defended what we now call a perfectionist political theory. This understanding of politics begins with some exemplary model of the good life and the good human. We saw how *energeia* was central to defining the excellent or virtuous life. A perfectionist political theory then uses that ideal model to design a society to promote the best kinds of lives.

In the beginning, with thinkers like Wilhelm von Humboldt, modern liberal democratic theory retained this notion of an ideal human being. This pattern of a good person informed early liberal ideas about a government that would promote self-realization. Liberalism, however, outlasted the notion of a common ideal and came to defend the goal of self-realization independent of any shared sense of what direction or shape a project of self-realization should take. John Stuart Mill's "harm principle" would enshrine this new kind of politics: do whatever you want, be whoever you want to be, so long as you don't harm others. The perfect form or shape of our *energeia* (the virtuous person) was replaced by the formlessness of volts.

The twentieth-century liberal political theorist Ronald Dworkin built on this tradition. He argued that since citizens differ in their conceptions of the good life, the government only treats them as equals if it is neutral on matters of ultimate concern. For him, this is not a retreat from moral issues but is in fact the "constitutive political morality" of liberalism (Dworkin 1978, p. 127). Our ruling morality, in short, just is neutrality. To each their own. As with energy, so with politics, being neutral is good. Neutrality means leaving the question of the good life open—to be answered by individual, not collective, choice.

Liberal democracy, then, maintains that the state should provide rights plus (to varying degrees) the economic arrangements and civil liberties to realize those rights. But it should neither promote the good nor justify its actions by an appeal to any particular conception of the good.

This neutrality thesis incorporates the economy and technology. The free market is a place to express whatever preferences you happen to have. Technologies are instruments to use as you see fit. Energy is woven across this trifecta of supposed neutrality: the economy, as regulated by democratic government, is just energy transformation, and energy is provided by technology and in turn powers our technologies from cell phones to tractors. It would seem that the state, the economy, technology, and E all participate in a happy coincidence of neutrality—supplying more and more means (a growing factor M) from which people can choose in order to give their very own answer to the open question: how best to live?

It is little wonder, then, that we think of energy as merely an instrument. It's like the proverbial hammer, which can be used to build a home or smash a man's skull. Electricity can be used to power a school or a prison, and so on. Energy, so this train of logic goes, is just calories—necessary but neutral. Just as a person could burn calories for any kind of pleasure or work, so a society can burn calories this way or that way. Moreover, whatever the end or application, it doesn't matter if the calories come from solar panels or coal plants. Like the hammer, energy does not dictate what will be done with it or what it will become. Rather, it enables its possessors to do whatever they want to do: more energy, more abilities.

*   *   *

There is truth to this orthodox view. But a moment's reflection reveals its limitations. High-energy ways of life differ from low-energy ways of life. Think about 1800 America versus today. The former is predominantly rural, handcraft, slow-paced, place-bound, and communal. High-energy life is overwhelmingly urban, consumerist, global, media-saturated, fast-paced, and bureaucratized. Americans in 1800 consumed 70 gigajoules (GJ) annually. Today that figure is over 300 GJ. The added energy didn't just "sup up" otherwise unchanged lifestyles. Rather, life changed dramatically with dramatically increased energy inputs. The added energy did not solely determine the changes, but it certainly was not neutral either.

To get at this, it is useful to think of energy as a medium, rather than a tool or instrument. And as Marshall McLuhan (1964) quipped, "the medium is the message."

In 1800, it took six hours of labor to earn an hour of reading light from a tallow candle. Today, it takes less than half a second to earn an hour of reading light from an LED light bulb. That's a 43,000-fold efficiency gain. But of course that doesn't mean we have 43,000-fold more free time to do the same old things, because an electric light bulb is not just a more powerful candle. As electric light and electricity more generally took hold, they transformed our world in their image. Energy isn't a neutral tool, it is a *medium* and the medium is the message (or the pattern). Light bulbs did things candles couldn't do, took us in new directions, and we came to *need* them in ever-increasing quantities. Smil (2017) notes that the average Briton consumes 11,000 times more light now than in 1750.

Energy services and the technologies involved are carriers of cultural messages, new ways of life and patterns of social organization, and different kinds of personal character, desires, and subjectivities. An automobile is not just a larger bicycle. Hooking up to a natural gas utility for central heating is not just getting a more efficient stove. A motorized tractor is not just a bigger horse. There are ways of life that suit an 8-horsepower machine. Those ways of life disappear when the horsepower crosses a threshold. Illich (1983) noted that the "kind of space" constituted by motors and feet differs in qualitatively important ways. As motors grow, they pattern spaces in determinate, that is, *non-neutral* ways.

As energy increases, the character of life changes. More is not just more. More is different.

The question is whether it is different in a *particular way* or just randomly. As we add more dots to the painting, do they give rise to a pattern or just a jumble? If it is just a jumble, we could salvage the neutrality thesis by acknowledging that more energy brings about different ways of life—but it does so in a neutral profusion resulting from individual choices, choices that have literally been *empowered* with more energy. In other words, one could argue that the new values that emerge are so varied and diverse that the choice of how to live is still left open. Maybe choices are simply multiplied as more machines are developed, leading to Daniel Boorstin's utopian democracy, which is able "to give everything to everybody" (1975, p. 102). Maybe volts just add more and more dishes to an expanding buffet.

This would be consistent with Leslie White's theory of energy. Recall that he discussed energy as the E in the equation $E \times T = P$. Energy multiplied by technology gives us productivity. Together this equation summarizes factor M or the means of any given culture. White thought of factor M as neutral with respect to factor E (where the E here means 'ends'). A growing factor M, he argued, provides greater assurance and abundance for pursing factor E. But according to White and the neutrality thesis, a growing factor M *does not* shape factor E. Again, more energy for a bigger factor M simply means more dots: more choice and more security.

That, however, is not the whole truth. The dots do in fact give rise to an image. A growing factor M stamps society in the mold of a particular pattern that is neither obvious nor exclusively dominant. What appears like an unfolding bounty of choice and diversity at the level of the dots, looks far more like a homogenous mold when seen from a different point of view. As Illich (1992) noted, no matter where you travel, the landscape is increasingly recognizable: cooling towers, electricity lines, highways, mega-cities, and industrial farms. It is the view spread out before the eyes of the Beaver of Progress: a patina of consumer choice spackled atop the same basic pattern of life delivered to us by Amazon and a handful of behemoths. The monopolizing tendencies of advanced capitalism are just part of the spread of this deeper, homogenizing pattern.

What is this pattern? The philosopher Albert Borgmann (1984) calls it the "device paradigm." The picture he sketches is like the upside-down version of convenience. That is, he reveals the dark side of a process the orthodoxy pictures in purely good/neutral terms.

A growing factor M is characterized by greater levels of convenience. More commodities are convened at our fingertips: coffee, frozen meals, fresh water, information, entertainment, and so on. Take warmth as a paradigmatic commodity. Since Prometheus stole fire, humans have had warmth. But to gather and chop wood, kindle the flame, and huddle around the fire is clearly not as convenient as touching a button on the thermostat to bring warm air quickly, safely, easily, cheaply, and evenly through the HVAC (heating, ventilation, and air conditioning) system of a house. Achieving this level of convenience requires energy inputs large enough to break free of the limits of a given locale and its fuel sources. The heat in a New England house might come from natural gas drilled in Texas, refined in Louisiana, and delivered in steel pipes mined from Michigan and forged in Colorado.

This pattern of growing convenience is neither unconditionally good nor neutral. Convenience disburdens us, say, from collecting wood, but this comes at the price of dis-engagement. It might be a burden to stoke the wood-burning stove on a cold morning, but it is what Borgmann calls a "focal practice," something that allows for skillful engagement with the world. By contrast, the central heater is maintained by a certified technician and entirely hidden from view. The old stove gathers the family together, becoming a "focal thing." The central heating system is, by contrast, not a thing but a 'device' that provides no point of focus or gathering. High-energy societies are characterized by the erasure of focal things, places, and practices as the rule of convenience and devices orders everything in its image.

For Borgmann, a 'thing' remains part of its context and sustains practices that are both means and ends. Raw ingredients from the market, for example, allow the occasion for the making of a home-cooked meal, which is not merely a burdensome means to the end result of food. It is, rather, a focal practice. By contrast a 'device' like the central heating system or the frozen meal has been decontextualized or torn out of the world of engagement and practice. This is essential to convenience, which must de-world things in order to stitch them together in ways that make them readily available.

The dichotomy between means and ends, factor M and factor E, is the defining pattern of the energy orthodoxy. As its pattern spreads, people lead increasingly 'commodious' lives, to borrow Borgmann's terms. That is, they lead lives in a foreground of conveniences where commodities are made readily available. Meanwhile, the background of means and machinery fades further, not just from view, but from engagement and understanding.

The question is: are we really free to choose the commodious life? And is it the best kind of life?

*   *   *

When the Amish consider adopting a new technology, they deliberate about whether it is compatible with their way of life. But *technology is our way of life.* I don't mean technology as this or that machine or artifact, but technology as the pattern of convenience that just is the energy orthodoxy.

What this means is that our democratic deliberations take place within the parameters already established by the device paradigm. The device paradigm, Borgmann writes, is seldom offered as a choice—as a way of life we are asked to prefer over others. Rather it is offered as the *basis* for choice. We can choose our commodities, but we cannot choose the commodious way of life. We can easily choose our devices, but we cannot so readily choose things over devices.

The philosopher Langdon Winner (1986) provides a good illustration of this with the mechanical tomato harvester developed in the early 1960s at the University of California. In the first decade after its introduction, these "factories in the field" drastically reduced the number of tomato farms and led to the elimination of 32,000 jobs. Yet, yields per acre and total yields increased due to the efficiency and power of the new machines. A coalition of rural communities and small farms filed a lawsuit against the University of California, arguing that tax monies had been spent on a project that helped some interests to the detriment of others, especially small farmers. The university responded that to accept these charges would be to "eliminate all research with any potential practical application."

So, even though the tomato harvester was the product of a state institution, it was treated as an infra-political or neutral device. Cheaper tomatoes more conveniently harvested were not put up for deliberation. They were, rather, taken for granted as "practical applications" that simply enhanced consumer choice. The loss of small farms reliant on hand-picked (and not genetically altered) produce was not pictured as a violation of liberal democratic neutrality. Rather, it was packaged as progress. Indeed, to seriously defend that way of life is to cast one's self as *impractical* and as anti-technology, anti-progress, and romantic. This is just to explain in a different way how the orthodoxy maintains the Overton Window or the boundaries for what shall count as realistic, serious ideas. Or, it is to show how destructive and even violent change is recast as progress.

Winner calls this the "moral claims of practical necessity." The efficient production of ever-more convenience is a moral imperative that trumps all other considerations. Appeals to deliberative democracy, traditional rural practices, or a slower pace of life, for example, appear foolish and utopian. Even to point out the decidedly non-neutral impacts of government-sponsored technologies is met with a yawn: "So what?! This is development and economic growth!" Highways, high-speed internet, and cell phone towers are also government-sponsored projects that introduce massive social change under the guise of neutrality. We can all sympathize with a desire to live free of traffic jams and constant information overload—but no one can seriously object to the practical necessity of highways and mobile data! That wouldn't be an appeal for greater democracy; that would just be foolish romanticism and backward thinking.

Liberal democracy hopes that technology and modern energy services will provide the instrumental means necessary to help people choose their preferred way of life. The problem is, though, that factor M (energy-technology-productivity) delivers more than it promises. It promises to leave the question of the good life open, but it delivers instead a definitive answer to that question in the form of the device paradigm. Convenience is the answer pre-chosen for you—how convenient! Liberty and justice can be realized only within the pattern of high-energy and high-tech convenience.

Borgmann argues that John Rawls and other major theorists of democracy are blind to the device paradigm. They talk about the fair distribution of basic social and economic goods like transportation and communication infrastructure. But this is already to choose the device paradigm as the basis for choice. Such structures can be indifferent only as to the choice of commodities (you could drive a Chevy or a Ford, and you can use an Android or Apple), but they are far from neutral as to the choice of engagement with things versus consumption of devices. Freedom means helping everyone afford a car and a cell phone, not creating opportunities to be *free from* cars and cell phones. Justice means getting everyone access to highways and high-speed internet, not allowing for flourishing lives *free from* highways and the internet. Progress means standardized tomatoes more efficiently harvested, not the preservation of local varietals. In other words, the device paradigm functions as a radical monopoly.

The energy orthodoxy understands itself as liberation from pre-technological ways of life. Our proverbial fisherman, for example, is caught up in fate or necessity. He is bound by place and tradition to live out a narrowly constrained destiny. Along comes E to break the chains. A way of life called progress and development promises to expand opportunities without prejudicing conceptions of the good. One can pick up whatever life project one chooses.

Yet we cannot avoid collectively determining the character of our community and our lives. The device paradigm becomes a new kind of fate, though of course we don't see it that way. Our attention is trained on the expanding dots, the consumer choices made from *within* the terms dictated by the device paradigm. The fisherman too made all sorts of choices from within his fate. Somehow, though, we think our choices are broader and more genuine.

But are they? The orthodox life is characterized by growth and accelerating change wrought by innovation. Under such conditions, whatever one chooses is increasingly free from contextual ties. The furniture comes in a box from Ikea. It must be light and mobile like life in general. Everything is fluid. Nothing is solid. The style of life in the device paradigm is one of shallow encounters with isolated and mobile commodities convened from out of an opaque background of machinery. Jobs that were once "for life" are increasingly contingent. People feel dispossessed, harried, upended, and anxious.

The philosopher Zygmunt Bauman (2000) summarizes the key upside-down aspects of orthodox convenience well: fragmentation, discontinuity, and inconsequentiality. Life unfolds like a TikTok video feed with images and episodes emerging and fading away in a stream that has no internal coherence and is anchored neither to a solid past nor a comprehensible future. For most people, this experience is part of a pervasive contingency and insecurity as the global economy shifts under their feet. But even the affluent are too distracted and debilitated to settle in on their deeper and fuller aspirations. Their vital energies are spent flitting around like maniacal moths from one shiny object to the next.

What the orthodoxy calls fate, Borgmann (1992) calls reality. The control of reality is good, but only *up to a point*. He uses a ski slope in Montana to make his case. Trees are felled, the path is groomed, and artificial snow might even be made. The experience is far from 'natural' in any purist sense of the term. But there is still a "telling continuity" with the surrounding mountain landscape and a "commanding presence" to the mountain itself. Reality, or fate, still sets the terms. You really are *out there* in the woods on the top of a mountain in the cold.

Compare that, Borgmann says, to the 'ski-o-rama' in a Los Angeles mall. This is an immersive virtual reality experience that perfectly simulates every aspect of the Montana mountain: the cold wind, the bumps under your knees as you swoosh downhill, the smell of the pines, a glance of a hare darting away, and so on.

The virtual ski experience has the virtue of being far more convenient or available. It is safer, cheaper, less messy, and faster than the real deal. But it is also disposable—pull off the virtual reality headgear and you are instantly back in a mall. That is, the experience is de-worlded in a way characteristic of E or modern energy services. As a result, it is less genuine or authentic. Whereas it was enticing in prospect, it will in retrospect be disappointing. We crave the real, which means that at times we must let fate simply hold sway. Danger, discipline, travail, and hardship are inconvenient sources of ultimate meaning.

*   *   *

Borgmann, Bauman, and others don't focus on energy with their theories of modern life. But the modernity they describe is driven and shaped by massive flows of energy. In particular, the modern *self* that experiences liberty does so in a decontextualized way, as an atomistic unit torn from tradition, community, and place. The focal things and practices that once provided enduring orientation are blown up by the sheer power of modern energy. This de-worlding (or alienation) is the unavoidable, *non-neutral* work done by volts.

Of course, people have some scope of freedom to choose focal things and practices. To do so, however, requires the realization that limitations are actually vital for genuine freedom. As the commentator David Brooks (2019) wrote, it's "the chains you choose that set you free." You need some framework to establish higher and lower goods, to guide your limited life force and give it meaning and purpose. Many people do in fact hitch themselves to projects larger than their own selfish consumerism.

They devote their life to family, to community volunteerism, or to some other form of service. To do so, they must practice discipline, relinquish much that would be more convenient, and work against the pattern of devices.

Theorists of the orthodoxy seem mostly blind to this dark, upside-down aspect of convenience. The ecomodernists, for example, celebrate increasing urbanization with unfettered joy, as if not registering the genuine goods associated with non-urban ways of life and as if blind to the ugliness spread out under the eyes of the Beaver of Progress. Bill Gates similarly touts development with unqualified optimism. Alex Epstein thinks that even Americans could use more modern energy services, despite the fact that they have long ago crossed the threshold where more E fails to bring further happiness.

More security brings more insecurity. Consider for example the ransom hacks against city governments and other cyberattacks on energy infrastructure. The ship channel in Houston where much of the US oil refining capacity occurs both supplies the security of a high-energy world and sits there like a giant Achilles' heel just waiting for a hurricane to cause widespread chaos. In Texas politics, oil refineries are considered "critical infrastructure" and you can get tossed in jail just for protesting at them even if you cause no physical damage. Obviously, they are sites of insecurity as much as security. They serve the infrastructure of our built world but threaten the critical infrastructure of a livable climate.

Controlling fate is always also tempting fate. The insecurities introduced by climate change are starting to make leaders of the global financial system nervous. In 2019, even the US Federal Reserve Bank (usually reluctant to talk about climate risks) issued dire warnings about looming threats from increased extreme weather events, especially flooding. As the Federal Reserve put it, "a new abnormal" is coming (see Flavelle 2019).

One of the most provocative paradoxes relating to democracy and security is the rise of what some call the "pink police state" (see Poulos 2014). A pink police state secures consumer freedoms even at the price of sacrificing civil liberties. You are free to shop but not to protest. The crucial line is no longer between private and public (as the government is permitted to reach deep into private lives with, say, surveillance or mandatory health-monitoring wearable devices). Rather, the key axis of political life is that between health-safety and sickness-danger. A society obsessed with controlling fate wants the government to guarantee our safety at any cost. The pink police state "is the product of a culture where people have given up on governing themselves and their impulses and turn to technocrats to shield them from the consequences of inevitable bad behavior" (Gobry 2016). I wonder if the coronavirus pandemic will grease the wheels on a slide into the pink police state.

In another way, the transhumanist Julian Savulescu (2009) illustrates the essence of the orthodox problem. He asks: what's the difference between a child learning math the old-fashioned way (by studying hard) or by taking a smart pill? He is asking about the difference between a focal practice (studying) and a device (the smart pill). Similarly, what's the difference between learning gymnastics the old-fashioned way (by practicing hard) or by receiving genetic enhancements? There is only one important moral difference, he argues. The new forms of enhancement (pharmacology, genetic engineering) are faster, easier, and can be made more reliable. We have a moral imperative, he says, to make such drugs and genetic techniques cheap and safe so that *all children* can benefit from them. Then becoming smart and athletic will be, well, far more *convenient!*

The problem is that he gives no weight to the process of self-creation— to the value of *inconvenient* labor or focal practices. The pianist who trains for years experiences that aspect of their self in ways qualitatively very different from someone who just downloaded the Piano Genius program directly into their brain yesterday. It is the same point about the ski-o-rama versus the real deal.

This is an argument Karl Marx made in his 1844 Manuscripts—part of what makes advanced industrial capitalism so demoralizing is the way it robs people of the meaning that comes from seeing their own handiwork objectified in a final product (see Frodeman 2019). The rule of convenience forces everyone to be a cog in the machine that produces devices that may be convenient and titillating but ultimately dissatisfying. They cannot understand or identify with the world around them. Savulescu is pushing this alienation to its extreme by alienating us from ourselves. We won't even know how we became who we are. Are we to let E become the author of our inmost being?

We just aren't going to find what we are looking for with ever-more elaborate devices. It's not waiting for us at the bottom of another barrel of oil or at the end of a wire enlivened by another wind turbine. The energy we lack is spiritual, and this lack is only exacerbated the more we confuse it for a quest for more stuff, more security, and more convenience. Our labors are not only a burden; they are also how we make life meaningful. This is what Hegel meant by "the labor of the negative." Nietzsche captures the point with characteristic verve in *The Will to Power*: "To those human beings who are of any concern to me I wish suffering, desolation, sickness, ill-treatment, indignities… I have no pity for them, because I wish them the only thing that can prove today whether one is worth anything or not—that one endures" (Nietzsche 1901).

In a 1972 essay titled "Energy is Eternal Delight," the Buddhist poet Gary Snyder counsels us "to grow with less":

The longing for growth is not wrong. The nub of the problem now is how to flip over, as in jujitsu, the magnificent growth-energy of modern civilization into a non-acquisitive search for deeper knowledge of self and nature.

Snyder might agree that "energy independence" is a contradiction in terms. A high-energy life is necessarily a life made more dependent on inscrutable systems, runaway desires, faceless bureaucracies, and merciless markets. That we consider all of this the very stuff of freedom and independence is our paradox. There are richer ores to mine if we do this jujitsu of turning inward to cultivate greater self-awareness, creativity, and compassion.

The upside-down of the orthodoxy is an existential crisis. Borgmann understands this in terms similar to his mentor, Heidegger: nihilism or "the forgetfulness of being." As convenience holds sway, the hedonic energies of pleasure-seeking crowd out other forms. Although the private experience of pleasure can provide impetus to act, it cannot give consistency, meaning, and seriousness to a person's life. We give ourselves over to trivialities as our commitments weaken. The open space that liberal democracy hoped to hold becomes an emptiness or void when it comes to purposes and goals. Amplified means detract from thoughtfulness about ends. Life moves too fast to figure out what it is all about.

Alas, Snyder's poetic Buddhism or Borgmann's focal practices are not among the leading alternatives to the device paradigm. There are some calls for slowing and simplifying life. There is some recognition that not all problems are matters of efficiency and scale. Self-knowledge, compassion, the appreciation of art, the difficult work of genuine dialogue—devices distract from such things far more than enhance them. Many people realize this.

But such viewpoints remain fringe. Far more central is the rise of dark ideologies with their promise to provide a sense of meaning and place in the tribe, the nation, the race, or the *Volk*. Across the globe, populist leaders are revolting against the globalizing, destabilizing, and homogenizing power of E in the crudest and laziest way possible. They are offering simplistic narratives with scapegoats: the 'other' who has caused you to feel lost, homeless, and afraid. We have learned too much from E—to seek the easiest path, to idolize comfort. We want the security of certainties even if they must come from the mouths of brutes. In this way, liberal democracy births illiberalism and Enlightenment breeds its own dark age.

Bibliography

Bauman, Zygmunt. 2000. *Liquid Modernity*. Malden: Polity Press.

Boorstin, Daniel. 1975. *Democracy and Its Discontents: Reflections on Everyday America*. New York: Random House.

Borgmann, Albert. 1984. *Technology and the Character of Contemporary Life*. Chicago: University of Chicago Press.

———. 1992. *Crossing the Postmodern Divide*. Chicago: University of Chicago Press.

Brooks, David. 2019. Five Lies Our Culture Tells. *New York Times*, April 15. https://www.nytimes.com/2019/04/15/opinion/cultural-revolution-meritocracy.html

Dworkin, Ronald. 1978. Liberalism. In *Public and Private Morality*, ed. S. Hampshire, 113–143. Cambridge: Cambridge University Press.

Flavelle, Christopher. 2019. Bank Regulators Present a Dire Warning of Financial Risks from Climate Change. *New York Times*, October 17. https://www.nytimes.com/2019/10/17/climate/federal-reserve-climate-financial-risk.html

208   A. Briggle

Frodeman, Robert. 2019. *Transhumanism, Nature, and the Ends of Science.* New York: Routledge.

Gobry, Pascal-Emmanuel. 2016. This One Political Theory Explains both Donald Trump and Hillary Clinton. *The Week*, March 18. https://theweek.com/articles/612533/political-theory-explains-both-donald-trump-hillary-clinton

Illich, Ivan. 1983 [2013]. The Social Construction of Energy. In *Beyond Economics and Ecology: The Radical Thought of Ivan Illich*, ed. Sajay Samuel. London: Marion Boyars.

———. 1992. Needs. In *The Development Dictionary: A Guide to Knowledge as Power*, ed. Wolfgang Sachs. London: Zed Books.

McLuhan, Marshall. 1964. *Understanding Media: The Extensions of Man.* New York: McGraw Hill.

Nietzsche, Friedrich. 1901. *Der Wille zur Macht. (The Will to Power).* Leipzig: C.G. Naumann.

Paulos, James. 2014. Welcome to the Pink Police State: Regime Change in America. *The Federalist*, July 17. https://thefederalist.com/2014/07/17/welcome-to-the-pink-police-state-regime-change-in-america/

Savulescu, Julian. 2009. Genetic Interventions and the Ethics of Enhancement of Human Beings. In *Readings in the Philosophy of Technology*, ed. David Kaplan, 2nd ed. Lanham: Rowman & Littlefield.

Smil, Vaclav. 2017. *Energy and Civilization: A History.* Cambridge, MA: MIT Press.

Snyder, Gary. 1972. Energy is Eternal Delight. *New York Times*, January 12. https://www.nytimes.com/1972/01/12/archives/energy-is-eternal-delight.html

Winner, Langdon. 1986. *The Whale and the Reactor: A Search for Limits in an Age of High Technology.* Chicago: University of Chicago Press.

16

Magic, Machines, and Markets

The great hope for a quick and sweeping transition to renewable energy is wishful thinking
Vaclav Smil (*2014*)

The energy orthodoxy has a way of turning today's magic into the machines of tomorrow. It specializes in transforming the outlandish into the ho-hum (witness: electric lights). Indeed, *magic* and *machine* are related. From the late fourteenth century, 'magic' was defined as the "art of influencing events and producing marvels using hidden natural forces." It shares the same root as 'machine,' namely **magh-* "to be able, have power." By the 1670s the Oxford English Dictionary defined 'machine' as "a device made of moving parts for applying mechanical power." Machines did the work previously reserved for magic—they influenced the world and performed marvels by unlocking the energies hidden in nature. As the mechanistic world of volts began to dominate, machines displaced magic or perhaps better said, they started doing the work of magic.

Yet magic still haunts many discussions about energy (see Pielke 2009). We need to only think about the hype surrounding biofuels in Europe or

© The Author(s) 2021
A. Briggle, *Thinking Through Climate Change*, Palgrave Studies in the Future of Humanity and its Successors, https://doi.org/10.1007/978-3-030-53587-2_16

ethanol in the United States—massive energy projects that promised the moon but have arguably done more harm than good. Biofuels may well exacerbate climate change by leading to deforestation and ethanol's primary contribution seems to be the further growth of fossil-fueled industrial agriculture and entrenched pork barrel spending.

The stunning success of the energy orthodoxy in creating previously unimaginable new realities makes it difficult to know when it is talking about real solutions versus magical solutions. Yet that same success makes this distinction ever-more crucial. The more we come to rely on the energy orthodoxy and its commitments to innovation and growth, the more imperative it is that it can deliver practical solutions rather than snake oil. One of the problems of living in a world that we cannot understand is that we won't be able to tell when the experts and policy elite are peddling wishful (magical) thinking and when their seemingly outlandish claims are actually feasible. We risk falling for magic tricks, because we've gotten so used to the magic of machines!

This makes the defining task of our age—decarbonizing the global economy—even trickier than it might seem. Can we really just switch over to solar and wind power? Can energy markets really sort this out? When Bill Gates and the other prophets of green capitalism and ecomodernity take the stage, what are they preaching: science or magic?

The sociologist of science Robert K. Merton (1942) summed up the situation aptly. The pronouncements of scientists, say, about wave mechanics or an expanding universe "run counter to common sense" and "cannot be checked by the man-in-the-street." A pseudo-scientific authority figure can offer myths that will seem more plausible than what the scientists are saying. So:

> Partly as a result of scientific achievements, therefore, the population at large becomes susceptible to new mysticisms expressed in apparently scientific terms. The borrowed authority of science bestows prestige on the unscientific doctrine. (Merton 1942, p. 277)

At a climate strike on my campus at the University of North Texas, I got into a conversation with a man who believed climate change was a socialist hoax. He told me to educate myself: "Google it, man! All the information is out there!"

The problem gets worse when even our trusted leaders (if such a thing exists anymore!) don't know the difference between machines and magic. Many of them may be advancing projects, policies, and proposals that are just simply unrealistic. They may be doing this as a prank; trolls have successfully stormed many high castles of policy-making. But they might be doing it with all seriousness and honesty. For example, many politicians in the United States are proposing plans to decarbonize the economy at rates exceeding 11% annually. Typical decarbonization rates for economies in the developed world are closer to 3% (see Pielke 2019a). So, are these more ambitious proposals mere wishful thinking or could they really happen? Americans would have to eliminate over 15 million combustion-engine cars annually. It could happen, but is that a "pigs could fly" kind of 'could'?

If we are to believe the climate science community, it's not a question of whether rapid, steep decarbonization *could* happen but a statement: that *must* happen. A 2018 IPCC Special Report (revised in 2019) vetted by over 1000 scientists concluded that staying at or below 1.5 °C requires cutting global greenhouse gas emissions 45% below 2010 levels by 2030 and reaching net zero by 2050 (IPCC 2018). Complete decarbonization of the economy in thirty years: that sure seems like magical thinking when energy-related CO_2 emissions were still increasing in 2018 and 2019.

(By May 2020, the economic shutdown precipitated by the coronavirus pandemic had caused greenhouse gas emissions to drop by 17% compared to the same period in 2019. Estimates for 2020 at that time forecast a total annual decline of about 7% (assuming activity picks up through the latter half of the year). If true, then even the massive economic downturn caused by the pandemic doesn't come close to the annual reductions called for by the climate science community.)

In 2018, the world consumed 11,865 million tons of oil equivalent or mtoe (see BP 2019). That year, global civilization added 280 mtoe of fossil fuel consumption and 106 mtoe of carbon-free energy consumption. To reach the IPCC 2050 target, these numbers need to flip dramatically—not only would all this new energy need to be carbon-free but existing fossil fuel energy would need to be decommissioned. The deployment rate of carbon-free energy needs to increase by about 800%. This is the equivalent of adding at least one new nuclear power plant or 1500 wind turbines *every day* for thirty years (see Pielke 2019c).

To call this a daunting task is an understatement. We can put the difficulty another way. In 2018, there were 1.7 trillion barrels of proved reserves of oil globally (that's 600 billion *more* barrels than 1998) (BP 2019). Are we really going to leave some 85% of that in the ground when it is the lifeblood of our civilization and the profits of the most powerful corporations in the world?

The IPCC report actually assumes that we are not going to do that. It relies on "negative emissions" or carbon dioxide removal (CDR) for most of its projected pathways to the decarbonizing targets. Basically, we are not going to be able to stop burning fossil fuels by 2050, so we will need machines to pull the carbon emitted right back out of the atmosphere. Metaphorically, this is like assuming we won't be able to resist eating the cake, so we are planning on liposuction surgery. CDR technologies do not currently exist at the scales or prices needed to even remotely get this job done. That doesn't necessarily mean that we are relying on magic, but we might be. Gates is banking that a new Edison/Newton/Tesla/Watt is going to save us. Like many faiths, the orthodoxy depends on a savior mythology.

The scale and speed of the global energy transition away from fossil fuels does indeed defy belief. Maybe we can pull it off. After all, massive energy transitions have happened in the past. Yet as Vaclav Smil has repeatedly shown, energy transitions have never been as rapid as we seem to imagine (e.g., Smil 2014). For example, consider the transition from wood to coal. The United Kingdom crossed from 50% wood power to over 50% coal in 1640. France didn't make this transition to coal until 1870. For the United States, it was 1884. Japan was 1901, Russia was 1925, and China was 1965. That means that the nineteenth century (and even in some ways the twentieth century) was still largely a time of wood.

Of course, the rate of innovation is growing today but as Smil (2013) notes, so too is the scale of the economy. So a jump in renewable energy from 3% of the energy mix to 8% is certainly significant. But that will be 8% of a larger number (remember the imperative of economic growth or

Locke's *increase*), so renewables will continue to supplement (and not simply replace) fossil fuels for a long time to come. Smil acknowledges that the political power of big fossil fuel corporations is a key reason for this. But physics, and not just politics, also explains the staying power of fossil fuels. They just happen to have really high-power densities, which is a measure of the amount of power (rate of energy transfer) per unit volume (see Smil 2015). Power density is perhaps the single most important determinant not just of energy systems but of modern civilization. A shift to less dense energy sources requires rethinking and restructuring energy systems, which takes time. The problem, though, is that time is not on our side.

In his latest book, *Growth* (2019), Smil once again voices the heterodox conclusion: the only way to really speed up the energy transition is to use less energy. Growth must end. The heretic! But he has good numbers—good enough to make us wonder if the orthodoxy is trying to prop up a magic show destined to be debunked. It wouldn't be the first time a major faith has toppled into the dustbin of history. Cultures have collapsed waiting for their saviors.

It is complex, because we do have some power to make the future through the force of our own beliefs. Part of any faith, including the energy orthodoxy, is what William James (1897) called "the will to believe." If we really *act* on our beliefs, we *could* will them into reality. Of course, that raises our earlier concerns about our rather anemic *willpower*. We might all *know* the numbers, but we don't really have the *will to believe* them. We can see those virtues in the young climate strikers and in others pushing for change, but is that enough?

*   *   *

As discussed earlier, decoupling is the idea of de-linking environmental harms from continued economic growth. At some point, peak impact is reached and then economic growth and per capita income gains continue even as environmental harms decline. A variety of factors can drive decoupling, including innovation, efficiency, and regulations. There are many ways to think about just what, exactly, is being decoupled. For example, economic growth can be decoupled from fossil fuels or it can be decoupled from energy consumption more generally (the latter case is measured in terms of energy intensity or the ratio of energy consumption and GDP, it is a matter of energy efficiency). Decoupling can be tracked in global terms or in specific sectors. For example, electricity use per household in the United States has declined since roughly 2012 despite big gains in household wealth over the same time period (see Davis 2017). The main reason for this is probably mandates for energy-efficient lighting, especially LED bulbs.

As we saw, the energy orthodoxy invests all of its faith in decoupling: growth must continue! The question is whether this is justified. We had better hope so, given that this is the linchpin of our entire way of life. Yet there are good reasons to be doubtful.

I'll start with an important paradox we neglected to mention in Part I: Jevons Paradox. The English economist William Stanley Jevons noticed in the mid-nineteenth century that increasing the efficiency of coal use (through improved technologies) actually increased consumption. This seemed paradoxical, because you would expect that more efficient technologies would utilize less of a resource. After all, efficiency means getting more output from less input. However (and this is where the market enters the equation), greater efficiencies lead to reduced prices, which tend to lead to greater demand and economic growth. In short, the conservation gains from greater efficiency are often eliminated by increased consumption. Economists call this the rebound effect (Fig. 16.1).

Smil (2010) notes that across the twentieth century, energy efficiencies tripled while global per capita consumption grew sevenfold. Efficiency gains have led to decreases in energy intensities, but per capita energy consumption continues to rise. Smil quotes Jevons arguing that his conclusion is still valid today:

> *It is wholly a confusion of ideas to suppose that the economical [efficient] use of fuels is equivalent to a diminished consumption. The very contrary is the truth.* As a rule, new modes of economy [i.e., efficiency] will lead to an increase of consumption according to a principle recognised in many parallel instances. (Jevons 1865, p. 140, emphasis in original)

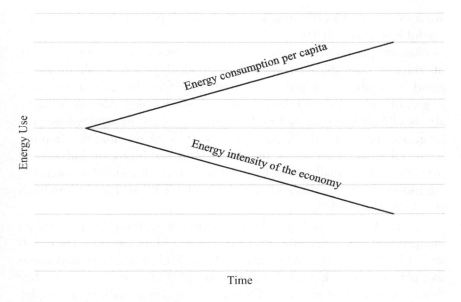

Fig. 16.1   The Shape of Jevons Paradox

In an advertisement for a fuel-efficient automobile, the narrator asks a happy customer: "What are you going to do with all the gas money you saved?" Her reply: "I'm going to buy plane tickets to Hawaii!" Again, we are driven to the root assumptions under the energy orthodoxy about boundless consumption as our ideal of the good life. If we don't question and find some way to constrain this way of life, then the Promethean quest of the orthodoxy will have a Sisyphean futility to it. Savings in one place will be plowed into consumption somewhere else. Decoupling starts to look like a game of whack-a-mole: save energy in one place only to have it pop up somewhere else as people spend their savings on more consumption.

But, you might ask, aren't renewable energy technologies going to save us? Again: maybe! It's hard to tell, because renewables can be so over-hyped that they start to appear like those snake oil cure-alls complete with no side effects. Things get really tricky here. For example, even the success of renewables can breed a "green paradox"—as fossil fuels start to lose market share, their prices will decline, which makes them more attractive to use. Decarbonization paradoxically gets harder the more successful it is (Pielke 2019b).

Now, in questioning renewables, we must be careful. There are plenty of critics of renewable energy who have an ideological or a political axe to grind. Most obviously, anyone invested in the fossil fuel industry is liable to give us a biased view. Tales of wind turbines killing birds, for example, are usually packaged in bad-faith framings that fail to acknowledge the far greater harms to wildlife posed by fossil fuels. Remember that framing is crucial when it comes to numbers: wind turbines may kill around 300,000 birds annually in the United States, but cell and radio towers kill 6.8 million birds and domestic cats kill between 1.4 and 3.7 billion birds (Erickson et al. 2014).

This is why I rely on Smil who is an equal-opportunity skeptic and an independent voice. When Mark Delucchi and Mark Jacobson (Jacobson and Delucchi 2011; Delucchi and Jacobson 2011) released a comprehensive plan to power global civilization on 100% renewable energy, they rightly drew tons of attention. Of course, the carbon industrial complex was dismissive of their analysis. I'll tune that out. But when Smil (2015) called their analysis "unbounded science and engineering fiction," well, I take that seriously. And Smil is not the only independent analyst who is critical about the dream of a total and rapid tech-fix to our energy crisis. The scholar Ozzie Zehner is a far cry from a fossil fuel hack, given his advocacy for walkable neighborhoods and simple living. Yet in *Green Illusions* (note the same trope about magic) he details the "dirty secrets of clean energy" (Zehner 2012). Solar panels, batteries, and wind turbines have a bevy of limitations and environmental side effects.

Of course, the orthodoxy's innovation engine might solve all these challenges. But the clock is ticking and "growth in energy use from fossil-fuel sources is still outpacing the rise of low-carbon sources and activities" (Jackson et al. 2018, p. 1). The orthodoxy's rosy picture of the coming Good Anthropocene has a serious contender—it is the picture of a global civilization that might have the winning formula for eventually getting to sustainability, but it is too little and too late. Developing countries are just growing too fast on the existing fossil fuel infrastructure and developed countries are just not doing enough. Ours might be one of those tragedies where the cure is discovered moments after the patient flat-lines.

This is one way to read the European Environmental Bureau's report *Decoupling Debunked: Evidence and Arguments against Green Growth as a Sole Strategy for Sustainability* (Parrique et al. 2019). The report reviews empirical and theoretical literature to evaluate the validity of the "decoupling hypothesis." The authors draw a conclusion that is "both overwhelmingly clear and sobering." There is "no empirical evidence supporting the existence of a decoupling of economic growth from environmental pressures on anywhere near the scale needed to deal with environmental breakdown" and "such decoupling appears unlikely to happen in the future."

The report acknowledges the importance of decoupling initiatives. Without them, things would be worse. But the point is that the primary or even exclusive focus by the orthodoxy on decoupling is a grave mistake. We need to broaden policy alternatives and open the Overton Window wider: "existing policy strategies aiming to increase efficiency have to be complemented by the pursuit of sufficiency, that is the direct downscaling of economic production in many sectors and parallel reduction of consumption." Once again that heterodox word bubbles up, that word rooted in the world of virtues: limit (see Latouche 2010). Those rare voices calling for degrowth might appear crazy today, but times change!

*   *   *

The word 'legerdemain' stems from the Middle French "light of hand." It means the quick and skillful use of one's hands when performing magic tricks. The so-called invisible hand of the capitalist market is the ultimate magician. Its top hat is that magic market mechanism: price. We see *price* everywhere—in advertisements, on labels, at the gas station pump, and so on. The invisible hand performs its *legerdemain* and hides so much inside of that magic hat! The moral imagination we are striving for—the ability to think CARBON upon waking up—requires us to see all the *hidden costs* behind the price tag. Those hidden costs are what economists call externalities. Like one of those endless magic scarves they are tucked into the top hat, enormous costs are somehow crammed into tiny prices. Can this illusion be sustained?

In chapter 9, we mentioned the wager between Julian Simon and Paul Ehrlich. The economist Simon won the bet, because five commodity metals were all cheaper after ten years despite the fact that they had all been intensively mined. Ehrlich expected that fewer metals in the ground would mean higher prices. But the opposite happened. Simon and the orthodoxy have an explanation for what seems like a paradox. 'Scarcity' just means a temporary increase in price. This spurs innovation, which either finds more of the resource or a superior replacement for it. Then price goes down, which means there is no more scarcity.

Yet, when the price is kept artificially low, no signal is sent. The innovation dynamo slumbers because no problem is detected. This is the case with carbon, which is too cheap.

As Simon noted, there is an awful lot that goes into making a price. The price of tin, for example, slumped in the 1980s because the world cartel backing its price went bankrupt (Worstall 2013). The price of tungsten tanked in the 1990s due to the collapse of the Soviet Union. Simon's point was that in the long run, such stochastic historical events are a wash. The real driver is innovation and that pushes prices down in ways that spur productivity and wealth.

There is, however, another "real driver" that pushes prices down, which is intimately related to Simon's neoliberal economics. Let's call it simply deregulation. This is a systemic feature of the modern economy and not some stochastic blip like a collapsing cartel or nation state. By 'deregulation' I mean a suite of magic tricks by which the true costs of energy and other commodities are hidden or made to disappear from the price. At the heart of this is, again, that primal energy source of labor power. Neoliberal capitalism of the kind championed by Simon is a global race to the bottom for cheap labor. The price of nickel or tin will decline if production shifts to areas with lax environmental and labor laws that pay miners dismal wages and externalize ecological and public health costs.

The extraction economy (fossil fuels as well as other 'earths') is increasingly propped up by various accounting tricks and financial wizardry designed to keep true costs out of the price. These tricks include tax breaks for oil and gas production or bailouts for coal mines. The energy analyst David Roberts (2019) broke down one model of externalizing costs that might best be called vulture capitalism. It entails a downward spiral of companies buying failing extraction businesses, restructuring to escape paying health, pension, and environmental obligations, then taking out massive loans, and finally declaring bankruptcy to walk away rich while communities (human and ecological) are stuck paying the bill for all the externalized costs.

Deficit spending is a key part of this system. The United States debt at this writing (and pre-pandemic stimulus measures) has topped $1 trillion. Fracking for oil and gas is now central to the global economy, yet it appears to operate by its own math-magic (Williams-Derry et al. 2019). Fracking companies borrow millions, over-produce, and enter the downward spiral of bankruptcy and buck-passing. Despite unprecedented production, many top companies continue to post negative cash flows, and those that show profits often can only do so through creative accounting techniques that omit key costs (see Mikulka 2018).

Another systemic driver of price is 'subsidy.' Indeed, subsidies are the most important types of magical accounting techniques. A 2019 International Monetary Fund study found that the United States spent $649 billion on subsidies for coal, oil, and gas in 2015 (Coady et al. 2019). That means that on average these subsidies cost every American over $2000 annually. These are the costs we pay that are not reflected in our air conditioning bill or the price at the gas pump. Many of these costs are wrapped up with the military industrial complex and its long history of interventions and entanglements in volatile areas that happen to sit atop massive deposits of oil and gas. Indeed, perpetual war conducted by the fossil-fueled military industrial complex is likely the single greatest *real-yet-hidden cost* of contemporary high-energy society.

Globally, the IMF put fossil fuel subsidies at $5.2 trillion in 2017. They use the term 'subsidy' to account for the "differences between actual consumer fuel prices and how much consumers would pay if prices reflected supply costs plus the taxes needed to reflect environmental costs" and other damages like healthcare costs. As one of my students put it in her dissertation: "what makes fossil fuels so cheap is all the money we're spending" (Rowland 2019). If we didn't spend all that money on subsidies, the IMF estimates a "net economic welfare gain" of $1.3 trillion. In other words, the powers that be are massively skewing the supposedly 'free' market toward net societal disutility.

Perhaps the most ominous subsidy is the one that treats the atmosphere as a free dump for greenhouse gasses. It would be much easier to think CARBON first thing in the morning if it had a price. Of course, it would be even easier if CARBON paid us. That is the idea behind carbon fee and dividend strategies like the one proposed by the Climate Leadership Council (2019) and endorsed by Ford, ExxonMobil, ConocoPhillips, and others. Their price for carbon: $40/ton. The money this generates will be distributed to Americans in a carbon dividend payment (about $2000 per family of four).

In a succinct formulation of the energy orthodoxy, the science policy scholar Roger Pielke (2019b) argues that any carbon tax needs to obey what he calls the "iron law" of climate change: when economic growth and emissions reductions collide, economic growth will win every time. So, we can internalize costs but we have to aim for the "Goldilocks Zone." If the carbon price is too low, we won't avoid the worst impacts of climate change. But if the carbon price is too high, it will threaten economic growth and therefore be politically untenable. Pielke sets his preferred starting carbon price at just $1/ton of carbon dioxide—enough to spur green innovation without slowing economic growth.

He may be right that a high price on carbon will be politically untenable. Yet what if carbon really does have a high price? I mean, what if—in order to survive—we have to reckon with costs that keep getting higher the longer we wish them away? We might be stuck between what is allowed by politics and what is required by physics. Is the iron law of economic growth bumping up against the steel laws of thermodynamics?

Prior to Adam Smith, the physiocrats ruled economics (see Nikiforuk 2012). They pictured the human economy as limited and governed by the productive forces of nature. Smith and, later, David Ricardo waved their magic wands to levitate the human economy above nature, as if all this unprecedented growth was fueled solely by the division of labor and free trade. Karl Marx and Friedrich Engels noted in their 1848 *Manifesto of the Communist Party* that the bourgeoisie had created "more massive and more colossal productive forces than have all preceding generations together." But they didn't question this magic trick—they just wanted the wealth more equitably distributed.

Maybe the physiocrats were right, though, and we just need to update the science behind their theory. We'd need to begin with the nineteenth-century physicist Rudolf Clausius, who formulated the first and second laws of thermodynamics: (1) energy can neither be created nor destroyed and (2) the amount of useful energy in a system inevitably declines. Clausius had a name for this loss of useful energy: entropy. At least, that's the word that stuck. Clausius actually preferred another term, *disgregation*, from *disgretatio*, roughly: things fall apart.

In the 1920s, the Noble Prize-winning chemist Frederick Soddy (1926) argued that the laws of thermodynamics govern all systems, including the human economy. Real wealth is based in energy, which is subject to entropy. By contrast, money and debt are "virtual wealth" which can keep growing. That levitating body of the human economy is just the virtual avatar of the real thing, which is rooted in finite natural resources that rot and decay and run out. The Romanian mathematician and economist Nicholas Georgescu-Roegen did more than anyone else to advance this train of thought with his 1971 *The Entropy Law and the Economic Process*. Georgescu-Roegen's work inspired the field of ecological economics through the work of Herman Daly.

Georgescu-Roegen's analyses (e.g., 1986) establish the test for the orthodoxy: can growth and innovation continue to happen within the constraints established by the laws of thermodynamics, especially entropy? Like Simon, the economist Robert Solow (1974) was optimistic on this score. But, also like Simon, he understood that it was an empirical question. We are indeed running the test right now. It might *seem like* all is going well as the global economy chugs ever upward. Yet this may be a castle built on an illusion that no amount of *legerdemain* can long sustain.

That, at any rate, is the conclusion of *Energy and the Wealth of Nations* (Hall and Klitgaard 2011), a reading of economics via the lens of ecology and physics. They show how the entropy of modern civilization is growing through calculations of energy return on investment (EROI). In a later publication with others, Charles Hall concludes: "The decline in EROI among major fossil fuels suggests that in the race between technological advances and depletion, depletion is winning" (Hall et al. 2014, p. 151). They then note that renewable energies do not offer "an easy solution" to keep economic growth churning. "It is probable," they argue that solutions will "have to come at least as much from an adjustment of society's aspirations for increased material affluence and an increase in willingness to share as from technology" (p. 151).

They are not optimistic about the prospects for such a revaluation of values from growing to sharing. I fear that they may be right when I consider how the honey badger managed to replace the canary.

Bibliography

BP World Energy Outlook. 2019. *BP Statistical Review of World Energy, 68th Edition.* https://www.bp.com/content/dam/bp/business-sites/en/global/corporate/pdfs/energy-economics/statistical-review/bp-stats-review-2019-full-report.pdf

Climate Leadership Council. 2019. The Four Pillars of Our Carbon Dividends Plan, September. https://clcouncil.org/our-plan/

Coady, David, et al. 2019. IMF Working Paper. Global Fossil Fuel Subsidies Remain Large: An Update Based on Country-Level Estimates. *International Monetary Fund.* https://www.imf.org/en/Publications/WP/Issues/2019/05/02/Global-Fossil-Fuel-Subsidies-Remain-Large-An-Update-Based-on-Country-Level-Estimates-46509

Davis, Lucas. 2017. Evidence of a Decline in Electricity Use by US Households. *Energy Institute at Haas,* May 8. https://energyathaas.wordpress.com/2017/05/08/evidence-of-a-decline-in-electricity-use-by-u-s-households/

Delucchi, Mark, and Mark Jacobson. 2011. Providing All Global Energy with Wind, Water, and Solar Power, Part II: Reliability, System, and Transmission Costs, and Policies. *Energy Policy* 39: 1170–1190.

Erickson, Wallace, et al. 2014. A Comprehensive Analysis of Small-Passerine Fatalities from Collision with Turbines at Wind Energy Facilities. *PLoS One* 9 (9): e107491. https://doi.org/10.1371/journal.pone.0107491.

Georgescu-Roegen, Nicholas. 1971. *The Entropy Law and the Economic Process.* Cambridge, MA: Harvard University Press.

———. 1986. The Entropy Law and the Economic Process in Retrospect. *Eastern Economic Journal* 12 (1): 3–25.

Hall, Charles, and Kent Klitgaard. 2011. *Energy and the Wealth of Nations: Understanding the Biophysical Economy.* London: Springer.

Hall, Charles, et al. 2014. EROI of Different Fuels and the Implications for Society. *Energy Policy* 64: 141–152.

IPCC. 2018. Global Warming of 1.5°C: An IPCC Special Report. Summary for Policymakers. https://www.ipcc.ch/site/assets/uploads/sites/2/2018/07/SR15_SPM_version_stand_alone_LR.pdf

Jackson, R.B., et al. 2018. Global Energy Growth Is Outpacing Decarbonization. *Environmental Research Letters* 13. https://doi.org/10.1088/1748-9326/aaf303.

Jacobson, Mark, and Mark Delucchi. 2011. Providing all Global Energy with Wind, Water, and Solar Power, Part I: Technologies, Energy Sources, Quantities and Areas of Infrastructure, and Materials. *Energy Policy* 37. https://doi.org/10.1016/j.enpol.2010.11.040.

James, William. 1897. *The Will to Believe and Other Essays in Popular Philosophy.* New York: Longmans Green and Co.

Jevons, William Stanley. 1865. *The Coal Question: An Inquiry Concerning the Progress of Our Nation, and the Probable Exhaustion of Our Coal Mines.* London: Macmillan and Co.

Latouche, Serge. 2010. *Farewell to Growth.* Trans. David Macey. Malden: Polity.

Marx, Karl, and Friedrich Engels. 1848. Manifesto of the Communist Party. *Marxists Internet Archive.* https://www.marxists.org/archive/marx/works/download/pdf/Manifesto.pdf

Merton, Robert K. 1942. Science and Technology in a Democratic Order. *Journal of Legal and Political Sociology* 1: 115–126.

Mikulka, Justin. 2018. Fracking in 2018: Another Year of Pretending to Make Money. *Desmog Blog,* December 18. https://www.desmogblog.com/2018/12/18/fracking-finances-record-oil-production-fuzzy-math

Nikiforuk, Andrew. 2012. *The Energy of Slaves: Oil and the New Servitude.* Vancouver: Greystone Books.

Parrique, T. et al. 2019. Decoupling Debunked: Evidence and Arguments against Green Growth as a Sole Strategy for Sustainability. *European Environmental Bureau.* https://eeb.org/library/decoupling-debunked/

Pielke, Roger. 2009. The Folly of 'Magical Solutions' for Targeting Carbon Emissions. *Yale Environment 360,* July 29. https://e360.yale.edu/features/the_folly_of_magical_solutions_for_targeting_carbon_emissions

———. 2019a. The Yawning Gap Between Climate Rhetoric and Climate Action. *Forbes,* September 19. https://www.forbes.com/sites/rogerpielke/2019/09/19/the-yawning-gap-between-climate-rhetoric-and-climate-action/#2e8de4ed2ec4

———. 2019b. The Case for a Goldilocks Carbon Tax. *Forbes,* September 13. https://www.forbes.com/sites/rogerpielke/2019/09/13/the-case-for-a-goldilocks-carbon-tax/#459bcee8f02a

————. 2019c. Net-Zero Carbon Dioxide Emissions by 2050 Requires a New Nuclear Power Plant Every Day. *Forbes*, September 30. https://www.forbes.com/sites/rogerpielke/2019/09/30/net-zero-carbon-dioxide-emissions-by-2050-requires-a-new-nuclear-power-plant-every-day/#e9faf635f7ec

Roberts, David. 2019. Coal Left Appalachia Devastated. Now It's Doing the Same to Wyoming. *Vox*, July 9. https://www.vox.com/energy-and-environment/2019/7/9/20684815/coal-wyoming-bankruptcy-blackjewel-appalachia

Rowland, Jennifer. 2019. Conceptual Barriers to Decarbonization in US Energy Policy. Unpublished dissertation, University of North Texas.

Smil, Vaclav. 2010. Science, Energy, Ethics, and Civilization. In *Visions of Discovery: New Light on Physics, Cosmology, and Consciousness*, ed. R.Y. Chiao, M.L. Cohen, A.J. Leggett, and C.L. Harper Jr., 709–729. Cambridge, MA: Cambridge University Press.

————. 2013. *Making the Modern World. Materials & Dematerialization.* Chichester: Wiley.

————. 2014. A Global Transition to Renewable Energy will take many Decades. *Scientific American*, January. https://www.scientificamerican.com/article/a-global-transition-to-renewable-energy-will-take-many-decades/

————. 2015. *Power Density: A Key to Understanding Energy Sources and Uses.* Cambridge, MA: MIT Press.

————. 2019. *Growth: From Microorganisms to Megacities.* Cambridge, MA: MIT Press.

Soddy, Frederick. 1926 [1933]. *Wealth, Virtual Wealth, and Debt: The Solution of the Economic Paradox.* London: Britons Publishing Company.

Solow, Robert. 1974. The Economics of Resources or the Resources of Economics. *The American Economic Review* 64 (2): 1–14.

Williams-Derry, Clark, et al. 2019. More Red Ink: Fracking Industry's Cash Flow Gap Widens. *Sightline Institute*, June 4. https://www.sightline.org/2019/06/04/fracking-industrys-cash-flow-gap-widens/

Worstall, Tim. 2013. A Little Detail about the Simon Ehrlich Bet. *Adam Smith Institute*, September 7. https://www.adamsmith.org/blog/economics/a-little-detail-about-the-simon-ehrlich-bet

Zehner, Ozzie. 2012. *Green Illusions: The Dirty Secrets of Clean Energy and the Future of Environmentalism.* Lincoln: University of Nebraska Press.

17

The Honey Badger in the Coal Mine

The honey badger don't give a shit.
Christopher Gordon (*2011*)

The canary is the bird of modernity. Colorful bangle, spoil of conquest and colony. Stolen from West Africa, the canary has long been known to Europeans as that happy prisoner singing gaily in his cage. Who would not be pleased to be rescued from the heart of darkness and given a domesticated, no, a *civilized*, existence? Child of Africa, you had been a flitting tatter in the green canopies, frightened at the slightest snap of a twig. And now you have a more secure perch. You have been given your freedom—freedom from want, from nature. Sing sweetly for the monks and courtesans. This is the story of modernity.

The coal mine is the canary's native habitat. 'Native' is of course an arbitrary designation, because Being is always in flux. The canary in flight and the canary in evolution are both the same: a smear, a rainbow blur. The bird of modernity has no essence, only existence. And existence moves, mangles, and mushes. When the Scottish scientist J.S. Haldane took you with him into the deep-dark of the coal mines, you had every

A. Briggle, *Thinking Through Climate Change*, Palgrave Studies in the Future of Humanity and its Successors, https://doi.org/10.1007/978-3-030-53587-2_17

right to be there. It was no more a violation of the natural order than your evolutionary departure from the dinosaurs. All is flux.

By the 1890s, when Haldane was sending bird cages down with miners, Britain was extracting 200 MT of coal annually and as a result it was ruling much of the world. Haldane applied the scientific mindset that is the hallmark of modernity and the energy orthodoxy. He saw the Earth as just one theater of universal actors. Such a view allowed him to see the bottom of the coal mine and the top of Pikes Peak as basically the same. There is that perfect similarity again. Mountain top and mine bottom were similar atmospheric extremes. Both presented certain challenges to human physiology. Both offered the canary a chance to play a leading role in the quest to conquer fate.

But it was the coal mine that presented the most pressing need. One-fifth of the male population of Great Britain worked in the mines. Over 1000 of them died annually in a cavalcade of disasters. To forge security from nature is itself a hazardous endeavor, exposing us to risks of our own making (see Beck 1992). At Haldane's time, the risk was not well-understood. What were the actual gasses causing sickness and death in the mines: whitedamp, chokedamp, blackdamp, after-damp? The imprecision of such terms infuriated the meticulous Haldane who performed tedious autopsies on the dead men perhaps all too soon for decorum but often too late for proper scientific analysis.

He also experimented on himself—inhaling carefully measured gasses in increasing dosages until he could no longer competently record notes and he stumbled about town, earning something of a reputation as a drunkard. Haldane's son would inherit this vocation of self-experimentation and blow out his ear drums in trials with atmospheric pressure. At cocktail parties, he is said to have blown rings of smoke through his ruptured ears with the nonchalance of a true philosopher.

The key property of the canary is its hyper-sensitivity. Because its threshold for toxins is relatively low, it serves as sentinel to cue the need for change before a point of no return is reached. Mice would also do, but "mice were no friends of miners" because they had a reputation for stealing sandwiches (see Goodman 2007). Besides, brightly colored birds were visible in the mine and also had the advantage of not just collapsing, like a mouse, but falling from their perches. The warning sign was much easier to see. Scientifically speaking, a bird is basically the same as a man—only with a faster metabolism. If the little man goes down, the large one will soon follow. A perfect similarity.

The large man will follow, that is, unless he adjusts his behavior. That is the key point. The bird of modernity is there to warn us, true, but not to stop us. It is not a totem animal symbolizing a taboo. When it falls, it is not a sign from the gods that we have over-reached our limits and that we must repent and cease. No, if it collapses, you exit the mine in an orderly fashion, fix the problem, and then proceed. This is the winning formula.

It is also the Enlightenment formulation for the relationship between reason (the canary) and passion (the miner). As David Hume, Haldane's fellow Scot and intellectual lodestar, wrote: "reason is the slave of the passions" (*Treatise of Human Nature* 1739–1740, II.3.3415). Passion, or desire, sets the agenda and provides the motive for action. But although the canary is indeed caged, it instructs. It is a sentinel, not the General. It does not educate in the sense of leading to better, more worthy desires. Rather, it only helps secure the rational and safe attainment of whatever desires happen to be in charge. The canary, like all modern reason, is a tool; it is instrumental.

The muscular miner and the delicate bird. This is the modern yang and yin. Eventually, the little feathered man lost his job to robots—in this case, the electronic gas detector, which is more reliable and more humane. (Then the coal miners started losing their jobs to robots, which are also more reliable and humane.) So, to update the yang and yin today, it is the smash-mouthed bulldozer and the intricate satellites above watching with hawk eyes. Writ large it is the brawny global GDP and the scrupulous scientific enterprise. The Tyne Valley mines were the beginning of the human remaking of the Earth. We are now all miners exposed to the technological risks of a planetwide operation to extract safety, comfort, and the good life. And we all depend on Haldane's intellectual heirs and their inscrutable scientific instruments—modern canaries—to warn us of impending problems.

The canary was the first pope of the energy orthodoxy and the coal mine was its cathedral. That we can keep our growing powers, our own creations, from turning around and destroying us (*pace Frankenstein*) is the central tenant of our faith. We are god-like, the orthodoxy tells us, but we must not misunderstand this. It does not mean that we quickly solve our problems and rest. No, like God the father, we are entangled in the happenings of our technological offspring—seeking to manage the consequences that we couldn't foresee. Having dominion over nature means being intimately attached to it, not being emancipated from it (see Latour 2012).

The ozone hole is a favorite example often used to justify the orthodox faith. The knowledge capacities of our satellite canaries grew in proportion to the technological capacities of our refrigerants. We were alerted to the problem and we reacted not by abandoning technology but rather by a further commitment to technology. The problem with this paradigmatic example, though, is that the same companies responsible for the problem had a technological fix ready at hand. In other words, the profit motive was never seriously in question. And, thus, there was no incentive to doubt the warnings given by the orbiting canaries.

As already noted, the same cannot be said of climate change and the grand challenge of decarbonization. The metaphor of a fallen canary itself falls flat. The canary is a poor stand-in for science, because the claims of science can be manifold and ambiguous in a way that a collapsed bird is not. Furthermore, the very success of the orthodoxy creates the conditions for us to lose the kind of vigilance demonstrated by Haldane and his canary. The further we get cozily ensconced in our technologically mediated worlds, the further removed we are from any signs of trouble. We start to feel immune and to act accordingly. So, both because the signals of science can be hard to read and because we are not all that motivated to pay attention, the canary model begins to break down.

*   *   *

We start losing the hyper-sensitivity of the canary. An animal of thicker skin becomes our totem. The honey badger is an aggressive, callous creature that first gained fame from a 2011 YouTube video. In the video, the narrator, Christopher Gordon, humorously notes that "the honey badger don't give a shit" as the animal smashes through bee hives and eats cobras all while suffering numerous stings and poisonous bites. The honey badger is the internet troll of the animal kingdom. It is heedless of consequences and oblivious to the concerns of others. The honey badger in the coal mine would keep right on pacing, growling, and chomping long after all the miners had collapsed from the after-damp.

This points to a problem for the energy orthodoxy. It is susceptible to being hijacked by a heterodox ideology that looks a lot like it, but is something else entirely, something very dangerous. To switch animal metaphors momentarily, it is like the way the viceroy mimics the monarch butterfly. This other ideology wears all the same trappings of the orthodoxy—it too celebrates the control of nature to improve the human condition. It too will talk about the importance of innovation. It also worships economic growth and free markets. It's just that it thinks the canary is a wimp and that science is a politically correct snowflake. This lookalike ideology also subscribes to the modern wager of no pain, no gain. It's just that it has a higher pain threshold. Indeed, it is numb. And in its unfeeling way, it rampages. It chomps through coal and other fossil fuels despite having thoroughly gorged itself and despite all the falling canaries, the warning signs, and dire predictions. You know, all that "fake news."

I am speaking primarily of the administration of US President Donald Trump and his agenda of America First Energy Dominance. If this was the only instance of this ideological mimicry, it would be worth discussing because of the out-sized influence of the United States on climate change. Americans represent only 5% of world population but consume 24% of total energy. Yet Trump's brand of head-in-the-sand nationalism has also taken hold in other parts of the world. That makes reflection on Energy Dominance all the more important.

The honey badger was the political symbol and spirit animal chosen by Stephen Bannon, Trump's former chief strategist. Bannon is the king of not giving a shit. He is a self-styled Leninist, meaning not just that he wants to tear down the state: "I want to bring everything crashing down and destroy all of today's establishment" (see Sebestyen 2017). Bannon is the former vice president of Cambridge Analytica, a data mining and political consulting firm. For the five years of its existence (2013–2018), Cambridge Analytica effectively ran around the world like a honey badger using big data and misinformation to chew through the democratic norms and civic fibers of several nations. It was forced to shut down in the wake of the 2016 US presidential election as it became apparent that the firm had harvested personal data from millions of Facebook users without their consent. Cambridge Analytica used the data to build psychographic profiles in order to manipulate voters in ways that some governments have characterized as "information warfare" (see Cadwalladr 2018).

Trump, despite the nuisance of an international consensus on climate change, has plowed ahead with a policy of Energy Dominance, which means extracting and exporting as many fossil fuels as possible (see Briggle and Sherrod 2019). Deregulation is key to Energy Dominance. Indeed, just two months into his administration, Trump issued Executive Order 13783, requiring all executive agencies to review any existing regulations "that potentially burden the development or use" of domestic sources of energy ('energy' is used synonymously with fossil fuels). The Environmental and Energy Law Program at Harvard University has since tracked the regulatory rollback efforts following from this order. The rollback tracker soon had dozens of entries on its list (Harvard Environmental Law 2019).

Another key to Energy Dominance is the federal budget. The Trump administration's fiscal 2019 budget proposal, for example, included: increases for oil production on the outer continental shelf, a 24% increase for fossil fuel research and development (including clean coal technologies), a 34% overall cut to the Environmental Protection Agency (EPA) and an 18% cut specifically to the EPA's enforcement division, the elimination of the Global Climate Change Initiative, elimination of five programs at NASA that monitor climate change impacts, and elimination of the Advanced Research Projects Agency – Energy (ARPA-E) Initiative, which funds high-risk research programs including grid-scale battery storage technologies to aid the transition to renewables. Trump's trade war with China also scuttled Bill Gates' Terrapower initiative, a next-generation nuclear power plant that held out the promise of decarbonizing the global electricity grid.

Yet even more essential to Energy Dominance is the honey badger's immunity to facts. This is accomplished by establishing an alternative universe of information. As energy journalist David Roberts has argued, Trump is the product of a right-wing media bubble that has created its own "tribal epistemology." For years, Rush Limbaugh, Fox News, and other right-wing thought-leaders have complained about the liberal bias of the media and other social institutions. What they wanted, though, was "not better, fairer, more scrupulous information referees" (Roberts 2017). They wanted "tribal information," that is, they wanted their own referees, their own rules, and their own truth regime. Every time Trump declares the media to be "fake news" or "the enemy of the people," he is reinforcing this tribalism. And with the infinity of information offered on the internet, there is plenty of room to build your own kingdom.

The result is a set of policies that are, like the honey badger, simply numb to reality. Some examples include a string of failed efforts to prop up what Trump calls "beautiful, clean coal" (Worland 2017). The Trump administration pulled out of the Paris Climate Agreement, despite unanimous international consensus. It killed the Clean Power Plan that had been the nation's signature climate policy. The Trump administration has insisted on cutting back fuel efficiency standards for automobiles despite opposition even from automakers. The administration has pushed for gutting requirements mandating the use of energy efficient light bulbs without any sensible justification. It continues to understaff environmental science agencies and cast doubt on climate science, despite the fact that even major oil and gas companies are endorsing carbon taxes and other climate mitigation strategies. Even the US military acknowledges the significant defense and national security threats posed by climate change (e.g., DOD 2019).

When automakers, the fossil fuel industry, and the military disagree with such an aggressive stance on fossil fuels, something disturbing is going on. Even young Republicans (18–34) are opposed to this kind of fossil fuel fixation (see Yoder 2019). This isn't the same-old disputes with the Sierra Club. It starts to look like Energy Dominance, unlike the energy orthodoxy, may not be the product of an epistemology at all. An epistemology implies a rationally defensible understanding of reality and a sincere quest for truth that abides by some standards of coherence and consistency. By contrast, Eric Levitz (2018) may be right that there simply is no rational policy agenda behind much of what the Trump Administration does. There is no good-faith, evidence-based argument that Trump actually won the popular vote in 2016, but nearly half of

Republicans believe that. Similarly, there is no rational way to defend the notion that climate change is a hoax perpetuated by the Chinese. This may simply be all about greed and not giving a shit.

If you don't want to know about problems, then ditch the canary and grab a honey badger. As Michael Lewis (2017) said in the context of an under-staffed and under-funded Department of Energy headed by a man who once called for its elimination:

> If you want to preserve your personal immunity to the hard problems, it's better never to really understand those problems. There is a downside to knowledge. It makes life messier. It makes it a bit more difficult for a person who wishes to shrink the world to a worldview.

This is reflected in Trump's remarks about the coronavirus in May 2020: "When you test, you have a case. When you test, you find something is wrong with people. If we didn't do any testing, we would have very few cases." Ditch the canary and the problem goes away. Brilliant!

* * *

Shrinking a complex world to a simple worldview is the way of the honey badger. Now, it is true that climate activists are also often guilty of this, especially with breezy narratives about simply flipping the switch from fossil fuels to wind turbines and solar panels. Simplifications by activists can lead to danger. For example, they can hound, censor, threaten, and silence expert voices who communicate a more nuanced view than the activists want to hear.

The far more dangerous threat, however, is coming out of the climate-denial camp behind agendas like Energy Dominance. Activists can give in to a false moral purity that can be censorious and counterproductive, but they generally lack access to power. It's the honey badgers in the halls of power that we really have to worry about. They have surrendered to the dark moral nihilism of the Joker. It's all a prank. And it is the fact that they have political and economic power that makes them the gravest threat to civilization.

The canary symbolizes sensitivity to nuance. The contrast between it and the honey badger runs much deeper than climate politics. Indeed, it constitutes a perennial theme in the history of ideas. It is the quarrel between Socrates and the sophists like Thrasymachus. Socrates is attuned to the many dimensions of justice, but in Book 1 of Plato's *Republic* Thrasymachus wants to boil it down to a simple answer: "justice is nothing other than the advantage of the stronger." Socrates, like the canary, is interested in finding out the truth. The sophists, those honey badgers, are interested in providing their paying clients with the tools to win arguments whether truth is on their side or not.

There is mimicry involved here too as sophists often pass themselves off as philosophers. Socrates, however, modeled a way of philosophy as offering primarily questions (he professed to only know that he did not know). The sophists, by contrast, profess to be experts and offer answers. Like a canary fallen from its perch and paralyzed, those who dialogued with Socrates often fell into a state of *aporia* or indecisiveness. Like the fleeing miners, they are compelled to rethink their practices. Those who counsel with sophists, by contrast, are given the tools they want to justify the continuation of whatever they are already doing. The gadfly Socrates pricked the conscience of Athens; the sophists offered soothing reassurance.

The leading sophist behind Energy Dominance is Alex Epstein, the self-styled 'philosopher' of energy and head of the Center for Industrial Progress. Earlier, I included his voice among the orthodoxy, but now I want to re-classify him as a mimic who only looks like the orthodoxy. He is the viceroy butterfly to the orthodoxy's monarch butterfly. He uses the same rhetoric of the orthodoxy but twists it into a defense of fossil fuels. Let's trace the way this contemporary Thrasymachus sells his rhetorical wares for "the advantage of the stronger," namely the fossil fuel industry (the criticism of Epstein below draws from Briggle 2018).

To do so, a bit of context is required. The administration of US President Ronald Reagan advanced environmental policies similar to the Trump administration. The Reagan administration zealously pursued deregulation and featured several leaders that were hostile to the very agencies they were heading. Julian Simon's work was influential in the Reagan administration in a way that Epstein's work is now widely circulated among thought-leaders in the Trump administration as well as right-wing energy policy circles (e.g., the Heartland Institute) more generally. The Simon-Ehrlich wager of the 1980s (discussed earlier) was, in some sense, replayed in a 2012 debate between Epstein and leading environmentalist Bill McKibben.

In many ways Epstein's book, *The Moral Case for Fossil Fuels* (2014), looks like Simon's argument. It's curious, then that Epstein hardly mentions Simon. I suspect that is because Epstein is actually playing Thrasymachus to Simon's Socrates. I don't want to exaggerate the philosophical nuances of Simon, but he was a sophisticated thinker. He was certainly committed to the free market and his own understanding of the dialectic between scarcity and abundance. But he was also an empiricist and, thus, open to revising his positions. He also recognized that policy debates cannot be reduced to numbers, because they are driven by values. Simon had a framing, to be sure, but he didn't boil everything down to a simplistic formula.

Epstein, by contrast, does just that. Despite all the similar trappings, he is actually selling a very different message. When Simon made his wager with Ehrlich, he was putting his money on a process. Epstein, though, is betting on a substance, namely, fossil fuels. Simon would never write a book making the "moral case" for any given resource. He was making the moral case for unlocking the ability of the human mind to nimbly shift from one resource to the next. That is the moral case for the orthodoxy. Simon championed the dialectic of human ingenuity and free markets in a way that was agnostic or neutral about any given resource. For example, people want cheap, reliable television in their homes. If copper cable becomes expensive, then switch to something else. People don't want copper. It is a mere means to get them the commodious, comfortable life championed by the orthodoxy.

Epstein, though, is making the moral case for a certain kind of resource, not a process that is agnostic about resources. He is aware of this. "Ultimately," he writes, "the moral case for fossil fuels is not about fossil fuels" (p. 34). That is a telling admission. Rather, "it's the moral case for using cheap, plentiful, reliable energy to amplify our abilities to make the world…a better place *for human beings*" (p. 34, emphasis in original). Epstein believes that "no other energy technology can even come close to producing that energy for the foreseeable future" (p. 34). But in saying that, he is aborting the imaginative impulse that is the beating heart of the orthodoxy.

By pinning his faith on a resource rather than a creative process, Epstein, the would-be disciple of Simon, betrays his master. Simon was a neoliberal (a free-market champion) and a technological optimist. Epstein wants us to think he is the same, but his allegiance to fossil fuels trips him up in two ways. First, he only celebrates technological creativity and markets when they favor his chosen fuels. And, second, he sets up all of his arguments to systematically discount the downsides of fossil fuels while hyping their positives. The result of these strategies is a book of half-truths. More generally, this is the consequence of the honey badger simplifications characterizing the Trump Administration's energy agenda.

As Simon noted, modern technology in a capitalist society is constantly evolving. The ends (commodities) will be provided through whatever means are cheapest and most efficient. Thus, because fossil fuels are mere means, they are vulnerable despite all their power. Witness the decline of coal, which is being partially cannibalized within the fossil fuel industry by natural gas. But it is also increasingly defeated on electricity markets by solar and wind as the price of these renewables continues to plummet dramatically.

As coal struggles to compete, the Trump administration abandons the orthodox logic of innovation and market signals to concoct various ways for the government to prop up their favored fuel. An administration that looks like the champion of free markets actually practices socialism for the wealthy by systematically favoring the fossil fuel industry. With his one-sided, resource-centered philosophy, Epstein provides moral cover for this approach. The cover gets gossamer thin with his extreme bias against renewable energy and his hand-waving dismissal of the IPCC and climate change. Simon's more orthodox logic of the eternal sunshine of spotless economic growth may be untenable, but at least his work was rooted in a coherent philosophy of dynamism. Epstein by contrast sings only anthems to the status quo. Simon's dialectic of scarcity and abundance pricks the conscience of even the most dominant market players, because their obsolescence might always be around the corner. Epstein sings a sweet lullaby to the forever power of the carbon industrial complex.

Epstein puts fossil fuels in front of a funhouse mirror that has the effect of ballooning their upsides and eliminating their downsides. Climate change, of course, is the elephant he makes disappear with some breezy cherry picking. As noted earlier, he argues that we are not taking a safe climate and making it dangerous through greenhouse gas emissions. Rather, we are taking a dangerous climate (prone to floods, fires, etc.) and making it safe by using fossil fuels to build shelters from the storms.

It's a decent point. After all, the aggregate global mortality attributed to extreme weather events dropped by over 90% since 1920, despite a four-fold rise in population (Goklany 2011). Yet this point can also be a diversion tactic. We are in fact doing both—increasing and decreasing risks. The IPCC continues to issue ever-more dire warnings and urgent calls to decarbonize the global economy. Epstein responds with a handful of charts and a reminder that CO_2 is, after all, *fertilizer*. I've been to Heartland Institute energy conferences where fossil fuel company executives wave Epstein's book in the air as if it were the Bible while they proclaim that "coal is green, coal is life!" This is what happens when a decent point about the benefits of fossil fuels in building resilient infrastructure is blown out of all proportion. Epstein's sophistry provides the flattering image the industry wants to hear. They can feel oh-so righteous about their "moral imperative" to export coal and natural gas to developing countries without any twinge of guilt or any second-guessing, because Epstein has found a way to eliminate climate change from the picture altogether.

This is the sophistry behind the Energy Dominance agenda that is helping to perpetuate fossil fuel dependence at a crucial moment when the opposite is needed. Epstein's sophistic simplification is a formula that could be used by, say, the sugar lobby: "Sugar is energy and energy is life! There are no downsides. Diabetes is a hoax." It's inane, but it sells. Sophistry is a political philosophy, because the assurance and clarity it produces are used to justify the behavior (and soothe the conscience) of the powers that be. To settle for half-truths is not to pursue wisdom, but it sure greases the wheels in the pursuit of profits.

Of course fossil fuels have been a tremendous benefit to humanity, but they also exact a steep and growing toll on the planet. Epstein only gives us one side of the story. Insofar as his *Moral Case for Fossil Fuels* is being used to justify similarly one-sided energy and climate policies, it amounts to an extraordinarily reckless book—indeed, a honey badger of a book.

Bibliography

Beck, Ulrich. 1992. *Risk Society: Towards a New Modernity*. London: Sage.

Briggle, Adam. 2018. Cherry-Picking Coal: A Review of the Moral Case for Fossil Fuels. In *Relations: Beyond Anthropocentrism*, ed., Giovanni Frigo. 6 (2), November, pp. 331–334.

Briggle, Adam, and Callie Sherrod. 2019. Postmodern Prometheus: A Discourse Analysis of Energy Dominance. *Sustainable Communities Review* 12 (1): 50–69.

Cadwalladr, Carole. 2018. 'I Made Steve Bannon's Psychological Warfare Tool:' Meet the Data War Whistleblower. *The Guardian*, March 18. https://www.theguardian.com/news/2018/mar/17/data-war-whistleblower-christopher-wylie-faceook-nix-bannon-trump

DOD. 2019. Report on the Effects of a Changing Climate to the Department of Defense, January. https://partner-mco-archive.s3.amazonaws.com/client_files/1547826612.pdf

Epstein, Alex. 2014. *The Moral Case for Fossil Fuels*. New York: Penguin.

Goklany, Indur. 2011. Wealth and Safety: The Amazing Decline in Deaths from Extreme Weather in an Era of Global Warming, 1900–2010. *Reason Foundation*. https://reason.org/wp-content/uploads/files/deaths_from_extreme_weather_1900_2010.pdf

Goodman, Martin. 2007. *Suffer and Survive: The Extreme Life of J.S. Haldane*. New York: Simon & Schuster.

Gordon, Christopher. 2011. The Honey Badger. *YouTube*. https://www.youtube.com/watch?v=qJF84oz93jw

Harvard Environmental Law. 2019. Regulatory Rollback Tracker. Accessed https://eelp.law.harvard.edu/regulatory-rollback-tracker/

Hume, David. 1739–1740. *Treatise of Human Nature*. Project Gutenberg. https://www.gutenberg.org/files/4705/4705-h/4705-h.htm

Latour, Bruno. 2012. Love your Monsters. *The Breakthrough Institute*, February 14. https://thebreakthrough.org/journal/issue-2/love-your-monsters

Levitz, Eric. 2018. Tribalism Isn't Our Democracy's Main Problem. The Conservative Movement Is. *New York Magazine*, October 21. http://nymag.com/intelligencer/2018/10/polarization-tribalism-the-conservative-movement-gop-threat-to-democracy.html

Lewis, Michael. 2017. Why the Scariest Nuclear Threat May Be Coming from Inside the White House. *Vanity Fair*, July 26. https://www.vanityfair.com/news/2017/07/department-of-energy-risks-michael-lewis

Plato. 2012. *The Republic*. New York: Penguin.

Roberts, David. 2017. Donald Trump and the Rise of Tribal Epistemology. *Vox*, May 19. https://www.vox.com/policy-and-politics/2017/3/22/14762030/donald-trump-tribal-epistemology

Sebestyen, Victor. 2017. Bannon Says He's a Leninist: That Could Explain the White House's New Tactics. *The Guardian*, February 6. https://www.theguardian.com/commentisfree/2017/feb/06/lenin-white-house-steve-bannon

Worland, Justin. 2017. President Trump Made a Confusing Reference to 'Clean Coal.' Here's What He Probably Meant. *Time*, August 23. https://time.com/4912730/donald-trump-clean-coal-phoenix/

Yoder, Kate. 2019. On Climate Change, Younger Republicans Now Sound Like Democrats. *Grist*, September 9. https://grist.org/article/on-climate-change-younger-republicans-now-sound-like-democrats/

18

Love, Death, and Carbon

We are in the beginning of a mass extinction, and all you can talk about is money and fairy tales of eternal economic growth. How dare you!
Greta Thunberg at the 2019 UN Climate Action Summit

It started when she was fifteen. Greta Thunberg spent the first three weeks of ninth grade on a school strike for climate. She sat outside the Swedish Riksdag (home to Sweden's legislature) demanding that her government reduce carbon emissions in accordance with the Paris Agreement. That was August of 2018. It had been Sweden's hottest summer on record, one that brought unprecedented wildfires across the arctic. At first, Greta was alone. Yet in a turn of events that could only happen in the age of social media, the beginning of the next school year witnessed nearly 1.5 million students participate in a global school strike for climate. Greta sailed across the Atlantic on a carbon-neutral boat to address the UN in New York City. She had gone from a protest-of-one to world famous in a year.

Her message to the UN was one of unbridled moral condemnation:

© The Author(s) 2021
A. Briggle, *Thinking Through Climate Change*, Palgrave Studies in the Future of Humanity and its Successors, https://doi.org/10.1007/978-3-030-53587-2_18

For more than 30 years, the science has been crystal clear. How dare you continue to look away and come here saying that you're doing enough, when the politics and solutions needed are still nowhere in sight. You say you hear us and that you understand the urgency. But no matter how sad and angry I am, I do not want to believe that. Because if you really understood the situation and still kept on failing to act, then you would be evil. And that I refuse to believe. (Thunberg 2019)

Though I was nowhere near as committed or articulate, I remember being a teenager and blaming the older generations for ruining the environment. They told me that they felt really bad about that, but they also said that the forests they felled and oil they drilled built our homes and fueled our dreams. They said they were acting out of love. I thought they were being selfish.

Yet somehow without noticing it, I have switched sides. I am the father who drives my son Max to gymnastics and buys my daughter Lulu a closet full of dresses. We have a dog and two cats; we feed them little brown pellets made of chicken or salmon and god knows what else. We've got a nice house in the suburbs. We drive the minivan to Colorado every summer to escape the Texas heat. My wife and I work hard to give them a good life full of experiences and opportunities. That's what parents around the globe are doing.

So, we have given the youth of the world an existential threat out of love. I guess I can only say: "You are welcome. I am sorry." We have a love problem. It is an intergenerational bear hug gone horribly awry. Or to paraphrase the rock band Aerosmith (sound track of my generation), our kind of love is the killin' kind. Because of the way we show our love, we may have to "learn to die" (Scranton 2015). That is, we may have to let go of an entire way of life and fashion new kinds of dreams. We have seen many superpositions: good and bad, power and weakness, rationality and insanity, why not love *and* death?

The school climate strike is an example of the public sphere as imagined by Hannah Arendt. Politics, she wrote, is about human plurality; it is the "sheer human togetherness" of marching down the street or sitting in protest at the Riksdag or the fracking site. The public sphere is the space that appears outside of labor, consumption, and work. There is little of that space left, because so much of what passes for politics today is really just the management of the economy. Our fundamentally political terms of *action* and *freedom* are understood almost exclusively in economic ways. Our freedoms are primarily of the consumer kind. The action that rules is the creative destruction of capital and market forces and the inscrutable dynamism of technological innovation.

And yet there are the students like monkey-wrenches in the machine. What's going to happen? Will they get ground down by the logics of growth and commercialization? Or is something new aborning? Arendt had a word for the new. She called it *natality*, and she said it is central to the human condition. Each new generation, each new person, is unique. Possibilities open.

Of course young people have long been rebellious and generations have always clashed. Yet never before has a generational rift been inscribed in planetary physics. A millennial born in 1990 has been alive for half of the carbon emissions ever emitted by humanity. To meet the Paris Climate Agreement goals, Lulu's cohort (born in 2012) has a lifetime CO_2 budget of 74 tons. Their parents' generation has a budget of 260 tons and their grandparents have 340 tons (Hausfather 2019a).

These numbers come from the climate analysis site Carbon Brief. Talking about the young climate strikers, their director put it bluntly, "Our analysis shows why they are right to feel so aggrieved and angry at the generations currently in power" (see Hausfather 2019b). Their parents and grandparents went on a carbon binge, leaving them with pennies left to spend and a ticking heat bomb in the atmosphere. When Max and Lulu's grandparents were born, atmospheric CO_2 concentrations were 310 ppm (parts per million). Now they are 415 ppm. The youth are trying to tell us that they are living on a different planet.

Though this message has inspired many, it has angered and frustrated others. Let's set aside the mockery and vicious attacks that have been leveled at Greta. Some things are beneath contempt. There is another critique, though, that is worth pondering. Aren't those young students being rather naïve and foolish? Isn't this all childish enthusiasm and idealism? Strikes make for good optics, but where's the substance? Where are the practical policy proposals? Americans, for example, are increasingly worried about climate change. Yet they aren't even willing to shell out $10 per month to fight the problem (Ekins 2019). We can't just wish away these kinds of political constraints.

This criticism paints the youth as *utopians*. Indeed, Thomas More's 1516 book *Utopia* is a touchstone for the tension between realists and idealists. The word *utopia* is a paradox—it can mean both 'nowhere' and 'a good place.' An impossibility *and* an ideal. In More's book, the central character Raphael has just returned to Europe from his travels to the island of Utopia where society is better. More, the idealist, implores Raphael to bend the ear of Europe's kings and persuade them to adopt the ways of the Utopians. Raphael, the realist, scoffs at the idea. For that to work, the kings would have to listen to ideas that they will consider to be totally outlandish such as limiting their wealth and power and abolishing private property. They won't consider this, because "Kings have no time for philosophy." They'd drum Raphael out of court, branding him a fool or a heretic. They might even have him beheaded (as More in fact was) for preaching dangerous, destabilizing ideas.

Unlike Raphael, Greta has stepped off the boat to stride into the halls of power. She looks world leaders in the eyes and tells them that their basic reality—their most fundamental dogma—is a "fairy tale." Economic growth cannot continue. That "unnatural growth of the natural" is eating her future. Her message is the logic of virtue, which is about thresholds and limits. She is attacking what Ivan Illich called "modern certainties" or what we have called the orthodoxy.

Yet we push forward with, in Greta's words, "business as usual and technical solutions." She sees this as a recipe for continued madness. And she diagnoses it as fundamentally a symptom of immaturity. We just want to keep entertaining ourselves. She notes how the IPCC projections rely on her generation sucking "hundreds of billions of tons of your CO_2 out of the air with technologies that barely exist." That's the magical thinking that we can expect from children, but these are the adults! Stone cold, Greta says: "you are still not mature enough to tell it like it is." Can't you see? The Emperor is naked! US Supreme Court Justice Anthony Kennedy wrote after a 2015 ruling: "The nature of injustice is that we may not always see it in our own times." Maybe, though, it's that we don't *want to* see it or we are *afraid to* see it.

*   *   *

In the *odium theologicum* or theological disputes during the Protestant Reformation, various factions accused each other of believing in fairy tales, magical spirits, and other baloney and gobbledygook. Something similar is afoot in climate politics, because orthodox thinkers from different backgrounds can turn the accusation of immaturity and wishful thinking right back around on the climate strikers and others preaching degrowth and radical change.

Roger Pielke is my former mentor and one of the voices I most respect in the world of climate policy. He is one of the most frequently cited experts in the peer-reviewed literature on extreme weather, and yet he has been unfairly demonized as a "climate change denier" (a patently false accusation). He has even been excommunicated by *Vox, The New York Times*, and *The Washington Post* in a way that damages discourse and unhelpfully narrows the range of expert voices brought to bear on our most complex global problem (see Pielke 2016).

Pielke makes a compelling case for the energy orthodoxy. He starts from the Kaya Identity where greenhouse gas emissions are pictured as the product of four factors: population, per-capita income, the energy intensity of the economy, and the carbon intensity of energy. Pielke (2010) further boils this down by combining the first two factors into GDP and the last two factors into the carbon intensity of GDP. That means there are just two levers for stabilizing the concentration of atmospheric carbon dioxide. We can either reduce GDP or reduce the carbon intensity of GDP. Shrink the economy or decarbonize it. Pielke (2012) says that there are only a "few brave/foolish souls who advocate a willful imposition of poverty." Indeed, *"the only option left is innovation in how we produce and consume energy"* (Pielke 2012, emphasis in original).

This hard-headed realism seems irrefutable when you look around at the built environment spread out under the eyes of the Beaver of Progress. Many of us live in a world premised on cars. We all live in a world powered by fossil fuels, from industrial agriculture to transport to plastics to electricity and more. We cannot turn that ocean liner on a dime or just wish it away. Degrowth sure seems like a recipe for economic collapse and all the pain, suffering, and starvation that would result. In addition to the built environment, there are entrenched cultures and values. For example, as Matt Taibbi (2019) colorfully put it: "The average American likes meat, sports, money, porn, cars, cartoons, and shopping." Moral outrage and green preaching are not going to change that. The orthodoxy may not always pack the moral punch of idealism, but it can claim the moral banner of pragmatism. We've got to play the cards we are dealt.

As noted earlier, Pielke often puts this in terms of the "iron law" of climate policy, roughly, any proposal that attempts to reduce GDP or curtail consumption is dead on arrival. For him, the "proper policy debate" is constrained by one question: "how do we stimulate energy innovation?" Anyone who talks about sufficiency or degrowth has not "done their homework." They are not proposing serious options. Instead, they "are invoking magic." I think Pielke's reply to the young climate strikers might be what he wrote in 2012: "Don't invoke magic, be informed."

Arthur Schopenhauer is rumored to have said, "All truth passes through three stages. First, it is ridiculed. Second, it is violently opposed. Third, it is accepted as being self-evident." Of course, Carl Sagan countered: "the fact that some geniuses were laughed at does not imply that all who are laughed at are geniuses. They laughed… at the Wright Brothers. But they also laughed at Bozo the Clown." So, who is genius and who is clown? Who is the realist and who is the idealist? A global ecological crisis tax on the wealthy and an upper limit to allowable wealth (Hickel and Kallis 2019; Robeyns 2019) seem like crazy ideas now, but the winds of change are blowing.

Before we toss peanuts at the young climate strikers, let's recall that our own orthodox faith was once ridiculed. Consider Niccolo Machiavelli's *The Prince*, written in 1517, just after More's *Utopia*. Machiavelli gave *virtù* a new meaning as the conquering of *fortuna*. Virtue came to signal a different kind of power—not self-control, but political conquest, innovation, and the rejection of authority. The ancients' project of trying to perfect the human soul, Machiavelli argued, was a fool's errand. It is better to take people as they are and design systems capable of withstanding, gratifying, and manipulating their passions. Remember, our snake brain (what some Buddhists call the monkey mind) is the one thing we are powerless to control!

This burgeoning modernity, which Descartes and Locke would later develop, seemed utterly ridiculous in the context of medieval feudal politics and Christendom. No one was talking about "the economy" at all, let alone imagining its endless growth. Heck, there were serious theological hang-ups about mining coal. Yet what was revolutionary on the lips of the early modern thinkers has become commonsense today. Obviously, any serious public discussion is limited to the gratification of given desires, not the shaping of better ones. That wasn't always so obvious. No regime in the time of More or Machiavelli was advancing a platform of innovation and economic growth—things that every politician nowadays takes for granted. Orthodoxies rise. They also fall.

*   *   *

The energy orthodoxy gives so much security to those of us in the developed world. Yet somehow we feel so precarious and vulnerable. Having been given our "vacation from humbler toil," we tune into the internet, which is designed to magnify insatiable desire, provoke rage, and spark fear, loneliness, anxiety, and feelings of inadequacy (see Lanier 2019). We lose perspective. This is an age-old problem about self-control and control of the world. The stoic Seneca wrote in Letter 47, "We are driven into a wild rage by our luxurious lives, so that whatever does not answer our whims, arouses our anger" (in Campbell 1969). We forget our strength— all that power at our fingertips—and instead we fixate on the things we cannot control. As a result, in the words of Seneca, we "grow white hot with rage" despite our "exalted station" in life. Or as stated in chapter 38 of the *Tao te Ching*

> The Master doesn't try to be powerful;
> thus he is truly powerful.
> The ordinary man keeps reaching for power;
> thus he never has enough. (Lao Tzu 1991)

It's never enough. The snake brain and monkey mind keep chattering.

Every summer, I attempt to hike one of the mountains in Colorado higher than 14,000 feet. They are known colloquially as the 14ers. For me, at least, this requires a lot of physical training. But it is not just about exercise. It is also a spiritual pilgrimage. Nothing centers and calms me like the experience at the top. With weakened legs, rolled ankles, and oxygen-starved lungs, I focus on the lunar rock face under my feet. One step at a time. Until I reach the ridge, when suddenly rock gives way to the open blue yawn of vast skies. The snow-capped San Juan's to the west, the hulking shoulders of Pikes Peak to the east, the razor teeth of the Sangre de Cristo Range to the south. I can sense the wholeness of Earth curving round, carrying me and mountains effortlessly through space. The alpine wind offers thin breaths of the thin ribbon of atmosphere where all living things find shelter from the awesome void overhead.

It is difficult to express those moments of profundity where we feel connected to the sacred. They are lost or at least diminished when we try to put them into words. In that time, you just *are*. You are vibrating as *energeia*, the activity of Being. "Give me a person," Epictetus once wrote, "who cares how he shall do anything, not for the obtaining of a thing, but who cares about his own energy" (in Nikiforuk 2012, p. 8).

Last summer, as I meditated in light-headedness atop Mt. Princeton, another hiker emerged at the top. He spoke loudly to a few others who had made it. There is good cheer among them as there should be. The experience, though draining, is also energizing. But this hiker crossed a line and profaned our sacred space and that holy, joyful moment. He pulled out his camera, hunched down, and started to record a YouTube video. He told his invisible audience where he was, and then he proceeded to deliver an extemporaneous lecture about electrical engineering.

Admittedly, he knew quite a lot about energy in the sense of volts and gates and amperes. But it struck me that he must know nothing about energy in another sense. The energy of Mt. Princeton was pulsing and radiating, but he couldn't tune into it. To do that, you must still yourself and quiet your mind. If you understand that energy, you know intuitively that you don't spend your time there yacking away into your phone. You don't look at your own face on the screen. You say little, you look long, and you simply allow the mountain to shine forth.

This book has been an exploration of these different energies of volts and virtues and how they have interacted to shape our world. It is a tight-rope walk. Books about energy fall into two categories. On one hand, there is the serious stuff of physics, engineering, and economics about energy systems, energy services, and energy policies. On the other hand, there is the 'softer' stuff about psychic energies, crystals, new age healing, and spirituality. But I think this is all wrong. There is a hidden spirituality behind all the techno-science and economics of energy, which is why I have used the faith-based language of the energy orthodoxy. That YouTuber atop the mountain epitomizes the spirit or religion of our age: always connected but never tuned in. His phone is his crystal. His own reflection caught in the screen under the mountain sun is his god.

I thought he crossed a line and acted inappropriately. But who am I to judge? In his lecture, he talked about electrical gates, which demarcate real lines that can be measured and controlled. In electrical engineering, you can send and stop pulses of energy across thresholds with great precision. The line I have in mind is more subtle and ambivalent. There are times when we must act and control, but also times when we must relinquish and be quiet. No one can say with precision when we cross those thresholds, but that does not mean that they don't exist.

I have tried to step carefully. It's not helpful to demonize technology. After all, the energy orthodoxy makes my mountain pilgrimages possible. The ongoing energy transition to a decarbonized economy depends on technological innovation. But I am with Greta on this one: maturity must also play a part in the making of a sustainable future.

The problem with the orthodoxy is its totalizing nature. It has part of the truth, but it overreaches to claim that it has the whole truth. Technological innovation cannot entirely substitute for wisdom. That's the lesson, anyway, that Zeus thought Prometheus had to learn. The orthodoxy is pushing Locke's doctrine of *increase* to an extreme. In the wake of religious wars, Locke understandably wanted to create a political system that made no claims on the virtues of citizens. Yet now the project of unbridled desire and limitless growth has led us into new kinds of dogmatic warfare and high-stakes insecurities.

The United States is the wealthiest nation on the planet with some of the best universities. Given historical carbon emissions, it has the most responsibility for tackling climate change. That the United States would elect a president who claimed climate change is a hoax is telling. This flight from reality, a defining element of immaturity, shows us how instrumental rationality can tip into irrationality. When we think that technological innovation is the *only* way to approach climate change, we shouldn't be surprised to find a clown show accompanies all the high-tech inventions: the joker superimposed with batman. Magic thinking superimposed with science. Protagoras told the myth of Prometheus in order to warn against this outcome. If we try to survive by fire alone—just technology—we wind up getting burned.

To think CARBON in the morning is like tuning into a metaphysical radio bringing us the voices of children seven generations in the future. Maybe we've already invented such a device. As the speed of the Great Acceleration ramps up, we compress futures more rapidly into the present day. It's like sound waves piling up in layers of increasing density at the nose of the plane as it approaches a sonic boom. The young climate strikers are prophets bringing us news from a future that we, with our high-energy spells, are superimposing atop the present.

The future is here and not here just like the walrus in your living room. My generation is left spinning like a qubit, loving and killing our children. Should Max and Lulu watch that scene of the walruses dying? Am I supposed to expose them to the world of our making or shield them from it? They might see themselves in those walruses. They too have been shoved, out of pure love, into the void. Netflix put a trigger warning in the documentary *Our Planet* to warn animal lovers about gruesome scenes. Jaguars kill an alligator, orcas maul penguins, and humans shove walruses off a cliff (remotely). With the exception of love, nothing is more natural and nothing is more upsetting than violent murder.

Bibliography

Campbell, Robin, ed. and trans. 1969. *Seneca: Letters from a Stoic*. London: Penguin Books.
Ekins, Emily. 2019. 68% of Americans Wouldn't Pay $10 a Month in Higher Electric Bills to Combat Climate Change. *Cato Institute*, March 8. https://www.cato.org/blog/68-americans-wouldnt-pay-10-month-higher-electric-bills-combat-climate-change

Hausfather, Zeke. 2019a. Analysis: How Much 'Carbon Budget' Is left to Limit Warming to 1.5C? *Carbon Brief*, September 4. https://www.carbonbrief.org/analysis-how-much-carbon-budget-is-left-to-limit-global-warming-to-1-5c

———. 2019b. Analysis: Why Children Must Emit Eight Times Less CO_2 than Their Grandparents. *Carbon Brief*, October 4. https://www.carbonbrief.org/analysis-why-children-must-emit-eight-times-less-co2-than-their-grandparents

Hickel, Jason, and Giorgos Kallis. 2019. Is Green Growth Possible? *New Political Economy*. https://doi.org/10.1080/13563467.2019.1598964.

Lanier, Jaron. 2019. Jaron Lanier Fixes the Internet. *The New York Times*. https://www.nytimes.com/interactive/2019/09/23/opinion/data-privacy-jaron-lanier.html

Lao Tzu. 1991. *Tao Te Ching*. Trans. Stephen Mitchell. New York: HarperCollins.

Machiavelli, Niccolò. 1517 (2000). *The Prince*, eds. Quentin Skinner and Russell Price. Cambridge: Cambridge University Press.

More, Thomas. 1516 (2016). *Utopia*, ed. and Trans. George M. Logan. Robert M. Adams. Cambridge: Cambridge University Press.

Nikiforuk, Andrew. 2012. *The Energy of Slaves: Oil and the New Servitude*. Vancouver: Greystone Books.

Pielke, Roger. 2010. *The Climate Fix: What Scientists and Politicians Won't Tell You about Global Warming*. New York: Basic Books.

———. 2012. A Primer on How to Avoid Magical Solutions in Climate Policy. *Roger Pielke Jr.'s Blog*, April 5. http://rogerpielkejr.blogspot.com/2012/04/primer-on-how-to-avoid-magical.html

———. 2016. My Unhappy Life as a Climate Heretic. *The Wall Street Journal*, December 2.

Robeyns, Ingrid. 2019. What, if Anything, Is Wrong with Extreme Wealth? *Journal of Human Development and Capabilities* 20 (3): 251–266.

Scranton, Roy. 2015. *Learning to Die in the Anthropocene: Reflections on the End of a Civilization*. San Francisco: City Lights Books.

Taibbi, Matt. 2019. Trump 2020: Be Very Afraid. *Rolling Stone*, August 19. https://www.rollingstone.com/politics/politics-features/taibbi-trump-2020-be-very-afraid-872299/

Thunberg, Greta. 2019. Remarks at the 2019 UN Climate Action Summit, September 23. https://www.npr.org/2019/09/23/763452863/transcript-greta-thunbergs-speech-at-the-u-n-climate-action-summit

19

Conclusion: Climate Change and the Future of Humanity

Nature, Mr. Allnut, is what we are put in this world to rise above
Rose Sayer (Katherine Hepburn) in "The African Queen," 1951

I started with an idea to write a book about climate change. I realized that energy would play a starring role. After all, so many stories about climate are about energy technologies: wind turbines, coal mines, next-gen nuclear power, floating platforms that harness wave energy, electric cars, and so on. That much was predictable.

Yet when I started digging deeper, it occurred to me that this is not just about energy, but energy understood in a new and peculiar way. The energy driving climate change is the E or volts of the orthodoxy—a new energy from a new faith in humanity. This universal currency breaks all boundaries and barriers. Indeed, the very concept of the 'climate' (like "labor power") is a category-crossing abstraction made possible by this sense of energy. We couldn't talk about and measure the climate until we could see energy flowing through the Earth's many spheres. E also jumps and jives across species in a way the old *energeia* never could. As Arendt (1958) puts it, we have introduced cosmic energies into earthly nature.

© The Author(s) 2021
A. Briggle, *Thinking Through Climate Change*, Palgrave Studies in the Future of Humanity and its Successors, https://doi.org/10.1007/978-3-030-53587-2_19

This new energy toppled an old world where everyone had their place and everything had its season. A good word for this is liberation. The new energy is about freedom.

So, what started as a book about climate change drove me through energy into political philosophy: the study, not just of power, but of the human condition. Climate change is the unintended consequence of a grand liberation scheme. As the transhumanist philosopher Steve Fuller (2019) puts it, liberalism or liberalization "is about your future not being overdetermined by your past" (p. 121). It began in the eighteenth century (just as E was being born) with men not being tied to the land or trade of their fathers. Slowly, it toppled many of the traditional sex roles assigned to women. Freedom means vacating the traditions, geography, culture, class, and even gender of one's birth. All of this liberation and transgression is the 'work' provided by E.

Fuller helped me to carry the logic of E and liberalism forward to its next stage: morphological freedom. Why not break free from the human form altogether? I mean "the human form" as in *Homo sapiens*, this neo-tenous ape. Fuller imagines all this energy helping us to blast off from our evolutionary trajectory. If our future need not be determined by our family's past or our culture's past, why should it be determined by our evolutionary past? We don't have to slog along from one ape to the next version of an ape. Remember the *tabula rasa* (blank slate) from Locke, that architect of liberalism: E can wipe the slate clean. We can radically morph. We can decarbonize and become silica, electrons, or qubits. We can upload our consciousness. We can become interplanetary. This is the shape-shifting logic of E. This is how a book about climate change opens questions about transhumanism and the future of humanity.

The economist Ted Nordhaus (2019) said that we are all neoliberals now, only some of us haven't realized it yet. Fuller takes this a step deeper: we are all transhumanists now, only most of us haven't realized it yet. I can put this in terms of the central problem with the heterodoxy, the approach grounded in virtues: where do you draw the line? Once the upward logic of liberation-E is in place, there is no non-arbitrary stopping point. Why wouldn't we mine other planets for energy resources? Why wouldn't we 3-D print replacement organs? Why wouldn't we keep going? We are on the path to a transhuman future, just most of us haven't caught up with this implication of the way we presently live. We draw lines today around a new normal only to transgress those outdated taboos tomorrow.

Of course, the heterodox voices may not simply be hypocrites (as implied by the framing that they are unwitting adherents to the orthodox transhumanism). Theirs is the precautionary approach. Rather than transhumanism, they subscribe to post-humanism. Transhumanists seek to indefinitely extend the qualities that most make us human, especially our intelligence. By contrast, post-humanists want to decenter and limit the human—they want to foreground the agency of non-human things. This is largely a matter of risk perception. In the case of climate change, the question is whether we think humans are (or can be) in control. One strength of the heterodoxy is its reminder that we aren't the only actors. Earth has her own energies, moods, and inclinations. We are oh so small.

Humility and hubris. The drama of climate change is both new and perennial.

The other strength of the heterodoxy comes from its attunement to the paradoxes of E. To be liberated is to become rootless and alienated. Moreover, the liberation afforded by E chains us to ever-greater needs and bigger systems with higher risks. The heterodox sentiment is epitomized in the concept of entropy: things do eventually wind down. Fate wins in the end. The shape of virtue (the golden mean) mirrors this diminishing return: things go up, then they come down. Finitude and mortality are the bitter in the bittersweet that is our earthly existence. To try to make it nothing but sweet is a fool's errand, a misunderstanding of the human condition.

The heterodoxy points out this foolishness when it comes to the so-called control of evolution. Even if we are to direct the forces of evolution in the name of freedom, the same ruling logic would steer all of our efforts. And that 'logic' is the most illogical kind: it is the logic of an arms race (see Ridley 2012). It is the Red Queen on her treadmill running faster and faster to stay in the same place. The host and the pathogen, the predator and the prey, the male and the female—they all engage in a dance back-and-forth with countervailing measures and natural selection pressures. It's a push and pull. All that energy to get nowhere! True freedom comes from stepping off the treadmill, not pushing the button for more and more speed.

To the orthodoxy, this will sound like giving up. Maybe entropy will rule the day, but aren't we here to test that proposition? Isn't that what it means to be human? And can you really look at the wonders produced by evolution and conclude that its logic is circular? We can't even imagine the wonders that a future AI will dream up.

At the heart of energy is paradox. The unchanging change. So, even entropy—the final heat death—must have its opposite, the never-ending life. Extropy means perpetual progress, and the perpetual overcoming of limitations (see More 2003). We are all extropians now, we just don't realize it yet. Entropy and extropy are two opposites superimposed atop one another like Schrödinger's cat, dead and alive. The idea of extropy was first conceived early in the history of volts as the "perpetual motion machine" that could function indefinitely without an external energy source, "because its interacting parts are subject to diminishing friction with each work cycle" (Fuller 2019, p. 154). It's the same logic as decoupling, or solving big problems that create smaller and smaller problems that are then solved.

It is all about efficiency gains. Of course, the first rule of energy states that there's no such thing as a free lunch! Maybe. But perhaps we can make lunch so cheap that it's practically free. If so, then we will harness E not only to solve climate change here on Earth but to solve the "climate problems" presented by other planets (e.g., insufficient oxygen or too much methane). We will terraform dozens of planets to serve us even better than good old Earth, home of humanity 1.0.

Still, though, the heterodox voices are nagging. The "most powerful nation on Earth" elected as president a reality television star, corrupt businessman, and former fake wrestling personality. President Trump has looked at climate change, the most daunting test for humanity, and decided it's a hoax. This is absurd, but it is accepted. That is a sure sign of a decadent society.

Decadence "refers to economic stagnation, institutional decay and cultural and intellectual exhaustion at a high level of material prosperity and technological development" (Douthat 2020). Our technical maturation comes at the cost of moral immaturity. All those easy calories make us fat and lazy. As the cultural critic Ross Douthat (2020) notes, the decadent society is "a victim of its own success." There's the paradox. Just as we harness E to build space-age technologies, we slump back into our snake brains. We are smart enough to tackle the many problems posed by climate change, but I worry that we just may not care enough.

While this book was under review, the coronavirus pandemic swallowed the world. Those with apocalyptic leanings have shifted their focus from climate change to Covid-19. The transhumanists among us might note with some glee about how we all are gambling like mad on a vaccine—that human enhancement technology that would resolve the drama as well as any deus ex machina. It's too early to know what the full impacts of the pandemic will be. Given that it is of historic proportions, however, readers will have it in mind. Indeed, every facet of life is filtered now through the shadow of the pandemic. I will leave most connections to the imaginations of the readers who will have the benefit of more experience than I presently have.

So, I will just pen a few provisional remarks—an odd way to end a book perhaps, but the only fitting way to keep a conversation going. I'll flag just five connections between the pandemic and climate change.

First, although pandemics are perennial, the growing human footprint on Earth and a deranged climate system are likely to increase future health risks. For example: more intense development and factory farming increase chances of species-hopping pathogens that can swiftly spread around a more interconnected globe. Of course, the building of a high-energy world also decreases health risks. For example: sanitation systems, transportation infrastructure, and modern communication technologies enable more people to have access to healthcare. Indeed, climate change is a matter of public health. It shows how high-energy civilization reduces *and* creates risks. The question, again, is whether we are smart enough to get out of the problems we keep putting ourselves into. Can we keep asserting our control and pushing back the hand of fate? Or, as Thoreau warned, will we get our foot caught in our own trap?

Second, nothing puts a dent in economic growth quite like a global pandemic: unemployment, consumption and production declines, and frozen supply lines. Never has spending money seemed so virtuous: ordering take-out or buying a gift certificate to keep local businesses afloat. We *need* to buy to keep a laboring society going. As I write this, airline travel is down 94%! Oil was briefly trading *in the negative* (meaning you had to pay someone to take it off your hands). Air pollution is down, water is cleaner, and wildlife is flourishing. Climate change activists hardly dared to dream of such things. The misery caused by economic contraction (including real threats of starvation) should make us wary of 'heterodox' perspectives that see degrowth as the solution to climate change. Yet, to be fair to these voices, what we are seeing with the pandemic is "negative growth" caused by a crisis, not planned degrowth. If we do have to shrink the economy to avoid a climate crisis, then we are clearly going to have to do that with great care so as not to cause more harm than good. As they say, the cure cannot be worse than the disease.

Third, unfortunately we are seeing some of the same epistemic tomfoolery with the pandemic and climate change. At least in the United States, there is the same politicization, obfuscation, denial, and fingerpointing. In Texas, for example, it would seem there are two pandemics (two realities) depending on your news source and political leanings. If you wear a face mask, you belong to one camp. If you don't wear a mask, you belong to the other. It's the same polarization that characterizes our insane climate politics. It is another marker of a decadent society.

Fourth, privilege and vulnerability look similar with climate change and the pandemic. The poor working class and people of color bear the brunt of both crises. Again, this is because both are matters of public health and access to the security supposedly provided by the high-energy way of life. It has been remarkable to see in the United States just how close so many people live to the margins. Behind the patina of affluence and abundance, there is very little reserve for most people. The unbelievable and unconscionable wealth disparities mark a crucial connection between the pandemic and climate change. Will we use this acute public health crisis to share the abundance of a high-energy society more equally?

This points to my last question: has the pandemic opened a window into a new way of life—perhaps something with a slower pace and fewer needs? And will it provide the occasion for something like the Green New Deal where people get back to work by building a decarbonized world? I hope so. The danger is that we will see a retrenchment of the status quo. Maybe the zeal to get things back to 'normal' (even though normal is wholly unsustainable) will be too strong. We'll set climate goals aside as luxuries we can't afford. We'll bail out big oil and gas. We'll prop up the airline industry. We'll stimulate consumer spending within the same petro-economy. We'll slash environmental and climate regulations to grease the wheels of economic recovery. In short, we'll blow the opportunity.

Only time will tell.

Bibliography

Arendt, Hannah. 1958. *The Human Condition*. Chicago: University of Chicago Press.

Douthat, Ross. 2020. The Age of Decadence. *New York Times*, February 7. https://www.nytimes.com/2020/02/07/opinion/sunday/western-society-decadence.html

Fuller, Steve. 2019. *Nietzschean Meditations: Untimely Thoughts at the Dawn of the Transhuman Era*. Basel: Schwabe Verlag.

More, Max. 2003. Principles of Extropy. https://web.archive.org/web/2013 1015142449/; http://extropy.org/principles.htm

Nordhaus, Ted. 2019. The Empty Radicalism of the Climate Apocalypse. *Issues in Science and Technology* xxxv (4). https://issues.org/the-empty-radicalism-of-the-climate-apocalypse/

Ridley, Matt. 2012. *The Red Queen: Sex and the Evolution of Human Nature*. New York: Harper Perennial.

Index

© The Author(s) 2021
A. Briggle, *Thinking Through Climate Change*, Palgrave Studies in the Future of
Humanity and its Successors, https://doi.org/10.1007/978-3-030-53587-2_19

Made in the USA
Las Vegas, NV
25 August 2023

76612247R00157